D1825660

Thoughts for Change
- We Can do it

THOUGHTS FOR CHANGE
-We Can do it

APJ Abdul Kalam
A Sivathanu Pillai

PENTAGON PRESS

THOUGHTS FOR CHANGE *-We Can do it*

ISBN 978-81-8274-707-4

First Published 2013

Reprinted in July 2013

Published by

PENTAGON PRESS

206, Peacock Lane, Shahpur Jat,
New Delhi-110049
Phones: 011-64706243, 26491568
Telefax: 011-26490600
email: rajan@pentagonpress.in
website: www.pentagonpress.in

Design and layout: Heena Sodhi

Printed at Aegean Offset Printers, Greater Noida

Dedicated to
the Youth of India
who have confidence in "I Can Do It",
that will lead to "We Can Do It",
resulting in "India Can Do It".

Royalty accrued to the authors through the sale of this book will be spent fully for societal missions

Contents

Preface

India is a unique Nation, constantly communicating the highest value of democracy and unity among diversity to the world. There may have been ripples here and there, time and again, but in an overall analysis **"India is India"**. When we talk about a nation and its greatness, everyone in that Nation has to be a responsible citizen. Every individual needs to create a value system and contribute individually and collectively for the nation to grow.

As the divine hymn says,

"Where there is righteousness in the heart
There is beauty in the character.
When there is beauty in the character,
There is harmony in the home.
When there is harmony in the home,
There is an order in the Nation.
When there is order in the Nation
There is peace in the World "

Spiritual wisdom and culture have been our strength. We survived as a Nation, the onslaughts of invaders for more than one thousand years and the numbing effects of colonization. We also learnt to adjust to rifts and divisions in our own society. But in the process of all adjustments, we also ended up lowering our aims and expectations and settled down to a lower mindset. From the great knowledge givers to the world and prosperity, we came down to the lowest strata of living at the time when we won our freedom. Eminent leaders emerged during the freedom movement, who tried to boost the spirits and minds of the people to recapture the greatness of Indians. It is time now that we must regain our heritage and wisdom to enrich our lives in the coming years. Presently, we are far from the expectations. Therefore, we need to evolve our own model of development, based on our inherent strengths to reiterate to the world that Indians are remarkable.

We need to set our targets high and work hard to achieve them unitedly. We may encounter genuine problems on the way but we have to defeat them so that the problems do not absorb us and become our masters. The country's economy steadily improved after the revival in 1991, however suddenly we seem to have lost control leading to a setback and economic slowdown. It is of course a temporary phase. If the countrymen resolve we will overcome this problem. Nothing is impossible if our aims are high and purposeful, and if we are determined to achieve them.

Mahatma Gandhi said:

> *"Men often become what they believe themselves to be.*
> *If I believe I cannot do something, it makes me incapable of doing it.*
> *But when I believe I can, then I acquire the ability to do it,*
> *even if I did not have it in the beginning."*

After India's independence, Pandit Jawaharlal Nehru structured the development through five year plans thereby systematically increasing the prosperity of the nation. He gave thrust for the development of Science & Technology through a policy resolution. From hand to mouth situation, we attained self-sufficiency in agricultural and milk production through mission mode programmes employing technology as a tool. Simultaneously, India gradually gained competence in the fields of heavy industries and power generation by Government funding and support from other countries. This created large scale employment in the country resulting in the development of skilled manpower. During the 1960s, the focus shifted to the development of critical technologies, especially, in nuclear energy, space and defence technology for self-reliance, pushing our country closer to the developed Nations. This vision has led India today to achieve the status of a nuclear weapon state and space club member that is self-reliant in defence technology. India's indigenous missile development programme for Defence through IGMDP encountered a huge road block through technology denial regimes, like MTCR, imposed by the developed countries on developing nations like India. The attempt was to contain the developing countries as buyers of second rated technological products. Our team broke the technology denial regime and worked on achieving self-sufficiency in missile development and production using a network of academic institutions, R&D organisations and Industries. Thus, we had succeeded in developing critical technologies for missile projects. Technology denied is Technology gained. Certain experiences in critical technology development are discussed in this book to give an insight into how we have done it.

India has seen significant progress in certain mission mode projects. In this pursuit, the human resources of India became the key factor to its growth, particularly in many niche technology areas. Advanced management practices are also important to keep pace with global growth in high technology areas. Consortium of institutions and matrix organisations provided excellence in high technology projects through synergy and cross flow of knowledge. Apart from indigenous development of products, collaborations and joint ventures helped in the fast development of certain critical technologies and the production of systems. This multi-pronged approach not only helped in acquiring key technologies in the shortest possible time but also opened up new frontiers in the development of new products leading to economic growth. Highly advanced supersonic missile, BrahMos was realised through a Joint Venture (JV) with Russian Federation. Governments, Administrators, Scientists, Technicians, Armed Forces of both India and Russia unitedly contributed their strengths and expertise for the success of the JV. It has become a unique role model to emulate and also to achieve product leadership, thereby giving competitiveness to India and Russia in the world.

All over the world, there has been considerable progress in technologies which are transforming human life. The technological progress like electronics and communication; advanced materials including composites, stealth and meta materials; aerospace technologies; energy devices; healthcare systems, biotechnology and so on. Sophistication in weaponry, development of accurate and multi role missile systems, sophisticated electronic gadgets for warfare, highly efficient ammunition and explosives, accurate and high performance small arms, amphibious vehicles, highly efficient radar and sonars, etc. vindicate the growth of science and technology in the defence sector.

New technologies are revolutionising research and development in biological sciences, chemical sciences and pharmaceuticals so that life-saving drugs are available at affordable prices. Recent development in nanoscience and technologies has found their use in every application especially in healthcare, environment, energy and defence. The interdisciplinary nature of nanotechnologies makes them versatile. The miniaturised nano devices and sensors will not only be used as targeted drug delivery systems, but will also be able to treat cardiac ailments, cancer, Parkinson's disease and many more critical disorders in the human body. Nanotechnology holds a lot of promise in the energy sector as well as in water resource management. Generation of clean and green energy should be the motto of scientists and technologists. The carbon nanotube based solar and fuel cells will create a new means of green energy at lower cost. Helium 3 (He-3) through fusion technology is expected to be a rich energy resource for the future replacing uranium through fission technology. There is enormous availability of He-3 on the Moon and Mars, and the prospective rewards would be stunning for those who embark upon this endeavour. The day is not far when mining industry is established on the moon so as to make the life of human beings better on earth.

These are the few examples of newer technologies that are going to revolutionise the future and empower mankind. India should utilize this opportunity in developing these technologies and to become a Developed Nation on par with industrially developed nations. But, many question whether India will become a developed nation? This is the result of an inferiority complex that was built over the years, when we were ruled by others. It is time that the youth of India must come out of this defeatist spirit and herald the message to the world that we are the knowledge power. We must shed the mindset of defeatism and energize ourselves with the spirit of confidence and victory with indomitable spirit to achieve our cherished dream of a developed nation for India".

Every one among the youth of India must prepare himself or herself to do remarkable work, which will create a place for him or her in the history of the world. History is not written for cowards and one who thinks small. If we think big, if we have courage, if we have a vision, if we do a great task or invention, or a great project, we find a page written in the history of the world.

Our dream is:

> *"I have no house, only open space*
> *Filled with truth, kindness, desire and dreams;*
> *Desire to see my country developed and great,*
> *Dreams to see happiness and peace abound"*

Saint Patanchali in his yoga sutra says:

> *"When you are inspired by some great purpose,*
> *some extraordinary work,*
> *all your thoughts break their bounds,*
> *your mind transcends limitations,*
> *your consciousness expands in every direction,*
> *and you will find yourself in a new great and wonderful world.*
> *Dormant forces, faculties and talents become alive,*
> *and you discover yourself to be a greater person,*
> *by far than you ever dreamt yourself to be."*

This book brings out the dynamism of the young in the advancement of science and technology in India and to highlight emerging opportunities that will place India among the top nations of the world. Advances in Technology lead to a quantum jump in the Economical status of the country. Industry developed countries have understood this fact. If we, too, understand it, then we will be one of the leaders in the world. The aim of this book is to bring out certain Thoughts for Change to kindle the fertile mind of the youth, to understand and make use of the game changer technologies that will eventually have great impact on the FUTURE INDIA, establishing that "We can do it".

APJ Abdul Kalam
A Sivathanu Pillai

Acknowledgement

This book is the outcome of many presentations, keynote addresses & lectures given by the authors in national and international seminars and other forums, visits to technological laboratories, discussions with experts during the last few years and also thoughts on Vision of Developed India. Authors would like to acknowledge and thank all the National and International organizations, scientific community for the information uploaded on their websites about their latest technological accomplishments, data, analysis which have been referred to in this book.

We have been fortunate to receive inputs from Government Ministries, Departments/organizations of Agriculture, Atomic Energy (DAE), Space (DOS), Defence Research and Development (DRDO), Earth Sciences, Bio Technology, IT & Telecommunication, New and Renewable Energy, Science and Technology (DST), Chemicals and Fertilizers, and from Universities, IITs, NITs and IISc on technological developments in the respective fields, and these inputs have been referred to.

Maj. Gen (Retd.) R Swaminathan, Mr. V Ponraj, Mr. Srijan Pal Singh, Mr. YS Rajan, Mr. D Narayanamurthy of ISRO, Mr. Dhanshyam Sharma, Mr. RK Prasad, Mr. H Sheridon, and also DRDO, ISRO, DAE and BrahMos teams including BrahMos Knowledge Centre, gave valuable suggestions and inputs. Our special thanks are due to Dr. Mayank Dwivedi, Mr. K Hariharan, Mr. Praveen Pathak and Mr. Rohan Mishra for their contribution in the preparation and shaping the inputs in the form of a book. All the contributions are of great value and we would like to acknowledge the help gratefully.

Finally, the authors would like to thank Pentagon Press, Mr. Rajan Arya and his editorial team for the corrections and for bringing out the book to the readers.

Introduction

"*Technology can effect the most fundamental changes in the ground rules of economic competitiveness and societal prosperity.*"

India has a long and distinguished tradition of science with great achievements from ancient times till date. Science and Technology in ancient and medieval India covered all major branches of human knowledge and activity, including mathematics, astronomy, physics, chemistry, medical science and surgery, fine arts etc. It was a land of sages, saints and seers as well as a land of scholars and scientists. The Binary Number System – the root of Computer Science, Decimal System, Pi and Zero as a number were invented in India. This was a great breakthrough in mathematics. The realization of Earth as a sphere by Varahamihira (Fifth century BC) and his calculation of the circumference of the earth at the equator as 40,350 km and the diameter as 12,845 km are very close to modern accepted data (circumference and equatorial diameter are, respectively, 40,075 and 12,756 km), are masterpieces of research in astronomy. Aryabhatta (Fifth century BC) developed the theory of earth rotation and Bhaskaracharya was the first to discover the gravity half a millennium before Sir Isaac Newton. This list goes on. Such was the contribution to Science by the ancient Indian scholars. Moreover, India contributed enormously to the origin and advancement of Civilization, the first one being Indus valley civilization.

During the great Mauryan dynasty under Chandragupta Maurya, 2300 years before, the Arthashastra, one of the greatest treatises on economics, politics, foreign affairs, administration, military arts, war, and religion was written by Chanakya, one of the ministers of the Empire. Every aspect of a great country was defined in detail. Significant emphasize was given to the knowledge and military supremacy. From such valuable documentation available to us, it is evident that India could again achieve a high level of supremacy with knowledge and technology as tools.

Today, one can easily realize that India has achieved significant success in varied fields of science and technology in the global arena. Government-sponsored scientific and technical research in diverse areas such as agriculture, biotechnology, pharma, industrial processes, communications, environment, nuclear energy, space, defence and IT & ITES have shown spectacular results. Now, India has expertise in the fields of astronomy and astrophysics, liquid crystals, condensed matter physics, molecular biology, virology, crystallography, applications software, cryptography, fast breeder and heavy water reactors, high energy materials and explosives, advanced composites, sensors, atmospheric re-entry technology, underwater systems and sensors, control and guidance of supersonic cruise missiles, robotics and autonomous systems, avionics, electronic warfare, lasers, and many more. The country also has embarked on research in niche technologies such as nanotechnology, fusion energy, photonics, artificial intelligence, network security, information warfare, hypersonic, high power microwave and so on. A large pool of scientific talent is emerging through the educated youth, bringing sustainability to the nation.

In the globalised world, competitiveness of the nation is the prime requirement. Technology is a tool to give a product leadership in terms of performance, quality, cost and availability in time for achieving competitive edge and hence economic superiority. Therefore, the major thrust is being given to increasing India's share in high-tech products, deriving value from technology-led exports. This will push India to be a major player in the emerging global economy. With its growing infrastructure and development of skilled human resource, India will touch new heights of sustainable growth in many sectors. Value based education and skill development for all the youth of our country is essential to achieve the higher level of economic standards and human resource index leading to top five nations in the world. Technology Vision 2020 evolved by 500 leading scientists of the country working for two years, outlines the road map to achieve the status of Developed India.

This book addresses ten unique technologies which are futuristic for competitiveness and global recognition. The ignited minds of the youth should ponder over these technologies and contribute to gain global leadership for our nation. These technologies answer to the futuristic needs of energy independence, availability of drinking water to all, green environment, affordable quality healthcare, security and strong defence for developed India to enable the people to live with happiness and prosperity, and also secured and safe.

Ten unique technologies discussed in the book are enumerated below:

1. Nano-Bio-Info Technologies and their convergence
   (Energy, Water, Healthcare, Defence)
2. Robotics, Artificial Intelligence and Cognitive Sciences
   (Automation, Industry, Defence, Space)
3. Sensor Technology – Photonics, Laser, MEMS
   (Surveillance, Security, Industry)
4. Materials Technology
   (Stealth, Smart, Composites, Meta-metals)
5. High Energetics
   (Explosives, Anti-matter, Thorium, Nutrino, Higgs-Boson)
6. Fusion Technology
   (Energy)
7. Space Technology
   (For Agriculture, Environment, Water)
8. Missile Technology
9. Hypersonics
   (Hypersonic Reusable Systems – Defence and Space)
10. Green Technologies

PART 1

Dynamics of Change

"Once a new technology rolls over you,
if you're not part of the steamroller,
you will be part of the road."

Stewart Brand

'Veda' means knowledge and the set of sacred books that originated in India. Vedic sciences are considered to be the richest and most comprehensive of ancient India. They include various disciplines of science like physics, mathematics, astronomy, cosmology, aviation, medicine, etc. Our ancient rishis (sages) used enormous mind power and possessed great scientific knowledge that they applied for the benefit of the community. The epics mention the power of the people. For example, in the Mahabharata (an Indian epic), King Dhridhrashtra wanted to know the happenings of the battlefield of Kurukshetra. Sanjay, the commentator, who had divine blessings visualized everything and narrated the complete happenings of the war to the king. In the Ramayana (another epic), it is believed that king Ravana had a Pushpaka Vimana like the present day air carriers. These are just a few examples out of several others. The unique spiritual and mind power, and intellectual thoughts portrayed in these epics in those ancient times amaze us today.

1.1 SCIENTIFIC INVENTIONS IN ANCIENT INDIA

Aryabhatta, Charaka, Sushruta, Panini were some of the eminent scientists of the ancient era. These scientists evolved many mathematical and scientific theories that can be proved using current methods. The saints and sages invented and discovered many new scientific breakthroughs. Aryabhatta (476 AD) was the first to proclaim that the earth is round and rotates on its axis. He is also acknowledged for calculating the value of π (Pi) to 3.1416 and sine table in trigonometry; Bhaskaracharya II (1114–1183 AD) was the first to discover gravity 500 years before Sir Isaac Newton; Acharya Kanad (600 BC) said "Every object of creation is made of atoms which in turn connect with each other to form molecules"; Acharya Sushrut

(600 BC) performed rhinoplasty (restoration of a damaged nose) and prescribed treatment for 12 types of fractures and six types of dislocations; Acharya Bhardwaj (800 BC) designed and described the techniques of aviation technology; and Acharya Kapil (3000 BC) outlined the concept of transformation of energy. The following paragraphs explain, in detail, some of the sages associated with scientific fields.

Astronomy and Mathematics

Born in 476 AD in Taregna (Bihar), *Aryabhatta's* intellectual brilliance remapped the boundaries of mathematics and astronomy. In 499 AD, at the age of twenty three, he wrote a text on astronomy and an unparalleled treatise on mathematics called 'Aryabhatiyam'. He formulated the process of calculating the movement of the planets and the time of eclipses. Aryabhatta was the first to proclaim that the earth is round, it rotates on its axis, orbits the sun and is suspended in space—1000 years before Copernicus published his heliocentric theory. He is also acknowledged for calculating the value of π (Pi) to four decimal places: 3.1416 and the sine table in

trigonometry. Centuries later, in 825 AD, the Arab mathematician, Mohammed Ibna Musa, credited the calculation of the value of Pi to the Indians, "This value has been given by the Hindus." Above all, Aryabhatta's most spectacular contribution was the concept of zero without which modern computer technology would have been non-existent. Aryabhatta was a colossus in the field of mathematics.

Astrology and Geography

Varahamihira (499–587 AD) was a renowned astrologer and astronomer, honoured with a special decoration and status as one of the nine gems in the court of King Vikramaditya in Avanti (Ujjain). Varahamihira's book 'Panchsiddhant' holds a prominent place in the realm of astronomy. He notes that the moon and planets are lustrous not because of their own light but because they reflect sunlight. In the 'Brihad Samhita' and 'Brihad Jatak', he revealed his discoveries in the domains of geography, constellations, science, botany and animal science. In his treatise on botanical science, Varahamihira presents cures for various diseases afflicting plants and trees. The sage scientist survives through his unique contributions to the science of astrology and astronomy.

Algebra and Mechanics

Born in the obscure village of Vijjadit (Jalgaon) in Maharashtra, *Bhaskaracharya's* work in Algebra, Arithmetic and Geometry catapulted him to fame and immortality. His renowned mathematical works called 'Lilavati' and 'Bijaganita' are considered to be unparalled and a lasting memorial to his profound intelligence. Their translation in several languages of the world bears testimony  to their eminence. In his treatise 'Siddhant Shiromani' he writes about planetary positions, eclipses, cosmography, mathematical techniques and astronomical equipment. In the 'Surya Siddhant' he makes a note on the force of gravity: "Objects fall on earth due to a force of attraction by the earth. Therefore, the earth, planets, constellations, moon, and sun are held in orbit due to this attraction." Bhaskaracharya was the first to discover gravity 500 years before Sir Isaac Newton. He was the champion among mathematicians of ancient and medieval India. His works fired the imagination of Persian and European scholars who, through research on his works, earned fame and popularity.

Atomic Theory

As the founder of 'Vaisheshik Darshan'—one of the six principal philosophies of India—*Acharya Kanad* (600 BC) was a genius in philosophy. He is believed to be born in Prabhas Kshetra near Dwarka in Gujarat. He was the pioneer expounder of realism, law of causation and atomic theory. He has classified all the objects of creation into nine elements, namely: earth, water, light, wind, ether, time, space, mind and soul. He says, "Every object of creation is made of atoms which in turn connect with each other to form molecules." His statement ushered in

the Atomic Theory for the first time ever in the world, nearly 2500 years before John Dalton. Kanad has also described the dimension and motion of atoms and their chemical reaction with each other. The eminent historian, T.N. Colebrook, has said, "Compared to the scientists of Europe, Kanad and other Indian scientists were the global masters of this field."

Chemical Science

Nagarjuna (100 AD) was an extraordinary wizard of science born in the nondescript village of Baluka in Madhya Pradesh. His dedicated research of 12 years produced discoveries and inventions in the faculties of chemistry and metallurgy. Textual masterpieces like 'Ras Ratnakar', 'Rashrudaya' and 'Rasendramangal' are his renowned contributions to the science of chemistry. Where the medieval alchemists of England failed, Nagarjuna had discovered the alchemy of transmuting base metals into gold. As the author of medical books like 'Arogyamanjari' and 'Yogasar', he also made significant contributions to the field of curative medicine. Because of his profound scholarliness and versatile knowledge, he was appointed as chancellor of the famous university of Nalanda. Nagarjuna's landmark discoveries impress and astonish scientists even today.

Medicine

Acharya Charak (600 BC) has been crowned as the Father of Medicine. His renowned work, the 'Charak Samhita', is considered an encyclopaedia of Ayurveda. His principles, diagnoses, and cures retain their potency and truth even after a couple of millennia. When the science of anatomy was confused with different theories in Europe, Acharya Charak revealed, through his innate genius and enquiries, the facts of human anatomy, embryology, pharmacology, blood circulation and diseases like diabetes, tuberculosis, heart disease, etc. In the 'Charak Samhita' he has described the medicinal qualities and functions of 100,000 herbal plants. He has emphasized the influence of diet and activity on mind and body and proved that the correlation of spirituality and physical health contributed greatly to diagnostic and curative sciences. He has also prescribed an ethical charter for medical practitioners two centuries prior to the Hippocratic oath. Through his genius and intuition, Acharya Charak made landmark contributions to Ayurveda. He will remain for ever etched in the annals of history as one of the greatest and noblest sage scientists.

Plastic Surgery

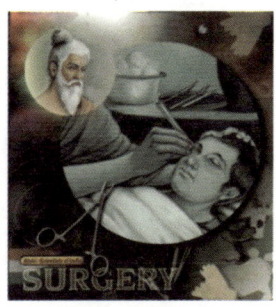

Acharya Sushrut (600 BC) is a genius glowingly recognized in the annals of medical science. Born to sage Vishwamitra, Acharya Sushrut details the first ever surgery procedures in 'Sushrut Samhita', a unique encyclopaedia of surgery. He is venerated as the father of plastic surgery and the science of anaesthesia. When surgery was in its infancy in Europe, Sushrut was performing rhinoplasty (restoration of a damaged nose) and other challenging operations. In the 'Sushrut Samhita', he prescribes treatment for twelve types of fractures and six types of dislocations. His details on human embryology are simply amazing. Sushrut used 125 types of surgical instruments including scalpels, lancets, needles, catheters and rectal speculums; mostly designed from the jaws of animals and birds. He has also described a number of stitching methods; the use of horse hair as thread and fibres of bark. In the 'Sushrut Samhita', he details 300 types of surgical procedures. This gives an enormous account of ancient Indians who were pioneers in amputation and caesarian and cranial surgeries. Acharya Sushrut was a giant in the arena of medical sciences.

Yoga

The science of Yoga is one of the several unique contributions of India to the world. It seeks to discover and realize the ultimate Reality through yogic practices. *Acharya Patanjali*, (200 BC) the founder, hailed from the district of Gonda (Ganara) in Uttar Pradesh. He prescribed the control of prana (life breath) as the means to control the body, mind and soul. This subsequently rewards one with good health and inner happiness. Acharya Patanjali's eighty-four yogic postures effectively enhance the efficiency of the respiratory, circulatory, nervous, digestive and endocrine systems and many other organs of the body. Yoga has eight limbs where Acharya Patanjali shows the attainment of the ultimate bliss of God through the disciplines of: yam, niyam, asan, pranayam, pratyahar, dhyan and dharma. Yoga has gained popularity because of its scientific approach and benefits. It also holds an honoured place as one of six streams in the Indian philosophical system. Acharya Patanjali will be remembered and revered forever as a pioneer in the science of self-discipline, happiness and self-realization.

Aviation Technology

Acharya Bharadwaj (800 BC) had a hermitage in the holy city of Prayag and was an ardent apostle of Ayurveda and mechanical sciences. He authored the 'Yantra Sarvasva' which includes astonishing and outstanding discoveries in aviation science, space science and flying machines. He has described three categories of flying machines: a) one that flies on earth from one place to another; b) one that travels from one planet to another; and c) one that travels from one universe to another. His designs and descriptions have impressed and amazed aviation engineers of today. His brilliance in aviation technology is further reflected through the techniques described by him:

a) *Profound Secret:* The technique to make a flying machine invisible through the application of sunlight and wind force.
b) *Living Secret:* The technique to make an invisible space machine visible through the application of electrical force.
c) *Secret of Eavesdropping:* The technique of listening to a conversation in another plane.
d) *Visual Secrets:* The technique to see what is happening inside another plane. Through his innovative and brilliant discoveries, Acharya Bharadwaj was recognized as the pioneer of aviation technology.

Cosmology

Celebrated as the founder of Sankhya philosophy, *Acharya Kapil* is believed to have been born in 3000 BC to the illustrious sage Kardam and Devhuti. He gifted the world with the Sankhya school of thought.

His pioneering work focused on the nature and principles of the ultimate soul (purusha), primal matter (prakruti) and creation. His concept of transformation of energy and his profound commentaries on atma, non-atma and the subtle elements of the cosmos place him in an elite class of master achievers—incomparable with the discoveries of other cosmologists. As per his assertion, prakruti with the inspiration of  purusha, is the mother of cosmic creation and all energies. He contributed a new chapter to the science of cosmology. Because of his extrasensory observations and revelations on the secrets of creation, he is known as the Father of Cosmology.

1.2 THE ERA OF TRANSFORMATION OF INDIA

India was one of the oldest centres of prehistoric culture of the world and was the cradle of one of the earliest rich and prosperous civilizations in history. The communities in ancient India were civilized and lived in planned cities with adequate facilities. They built houses of brick, wore cotton clothes and made beautiful gold and silver jewellery, pottery and toys. The heritage of India is the result of developments in the social, economic, cultural and political life of Indians over a period of thousands of years. The Indus Valley civilization and the ruins of Mohenjodaro and Harappa bear testimony to the fact that even as early as 2500 BC, India had skills to develop agriculture, a drainage system, well-planned streets, pottery, tools, jewellery and artefacts. The Harappan culture was the first urban culture to emerge in India. Many of its features distinguished it from all its contemporary cultures in other parts of the world, and made it distinctly Indian.

The Vedic Period followed and was marked by far-reaching changes in almost every facet of life in India. This period saw the spread of agriculture over large parts of the country, the surge of cities and the formation of states. During this period, both political unity and cultural unity received increased importance. The four main Vedas—Rig, Yajur, Sama and Atharva—relate the greatness of ancient Indian literary work. The Upanishads dealt with questions like the origin of the universe, birth and death, the material and spiritual world, nature of knowledge and many other questions. Another body of literature to grow in the early period was the Vedanta which, besides rituals, was concerned with astronomy, grammar and phonetics. One of the most outstanding works of this period was a classic on Sanskrit grammar, the Ashtadhyayi by Panini. All these works were in Sanskrit. They were handed down orally from generation to generation and were put in writing much later, because of which most of the work done by our great people could not be passed on to the next generations resulting in degradation of values. Meantime, religions also originated with the influence of beliefs and practices. The rise of cities, crafts and trade also furthered the process of cultural unity. Later, the Magadh empire around sixth century BC saw the birth of cities and use of coins. The first society established in the Indus civilisation became a model for the human race. The prosperity continued during the great Mauryan rule of Chandra Gupta Maurya and later in the third century BC under Emperor Ashoka the Great when India spread its rule far and wide. He unified almost the entire country under one empire but renounced the use of war as state policy. Instead, he declared the victory of righteousness as the real victory. In him, we also find a change in the ideal of kingship. India was aptly called the 'Jewel of the East'. By the time the ancient period of Indian history came to a close, India had developed a culture which was marked by features that have characterized it ever since. During the medieval period, some of the achievements of the ancient times were carried forward and new and magnificent structures were built on those foundations.

The metallurgical works of the dancing girl of Mohenjodaro, the earliest known Indian lost-wax process cast bronze figure (third millennium B.C.), the panchloka (five metals) idol of Lord Nataraja in cosmic dance (eighth century A.D.) using investment casting process at Chidambaram, Tamilnadu, the "rustless" 7 m iron pillar of Mauryan dynasty at Delhi were evidence of a high degree of technical excellence in shaping metals and alloys as a single system. It is a testimony to the skill of ancient Indian blacksmiths. The Grand Anaicut built by the Chola king Karikalan around the second century is an ancient dam built on the Kaveri River in Tiruchirapalli in the state of Tamil Nadu and is even today an engineering marvel. Ancient Indians also excelled in medicinal and martial arts. Bodhi Dharma, the third son of Pallava King of Kancheepuram in South India during fifth century went all the way to China to help people with his knowledge of herbal medicines for curing unknown dreaded diseases and with his knowledge of martial arts protected them from the tribal war. He taught martial arts to the Chinese through his teachings and established the famous Shaolin temple. Nalanda was the first university to attract scholars from more than thirty countries to spread the knowledge. Spread of the Indian kingdoms to the south east countries by the Chola kings Raja Raja and Rajendra enhanced the cultural spread to those countries. Such was the great contribution of India to arts, culture, religious thoughts, science (mathematics, astronomy, cosmology, atomic theory, medicine, yoga, and so on), technology (metallurgy, aviation, township, etc.), and civilisation that spread to the world.

Alas, more than one thousand years, India's culture, value system and wealth eroded due to continuous invasions by several kings and countries. Not only our wealth was looted, many of our archaeological monuments were also destroyed. Alongwith increased population and non-participation in the industrial revolution, the country slipped down on the economic scale. Figure 1.1 brings out India's prosperity dynamics curve.

The First War of Independence in 1857 saw the birth of the vision for an independent India, which was achieved through a non-violent movement, in 1947. This movement attracted the best of leaders in politics, public life, music, poetry, literature, and science. The

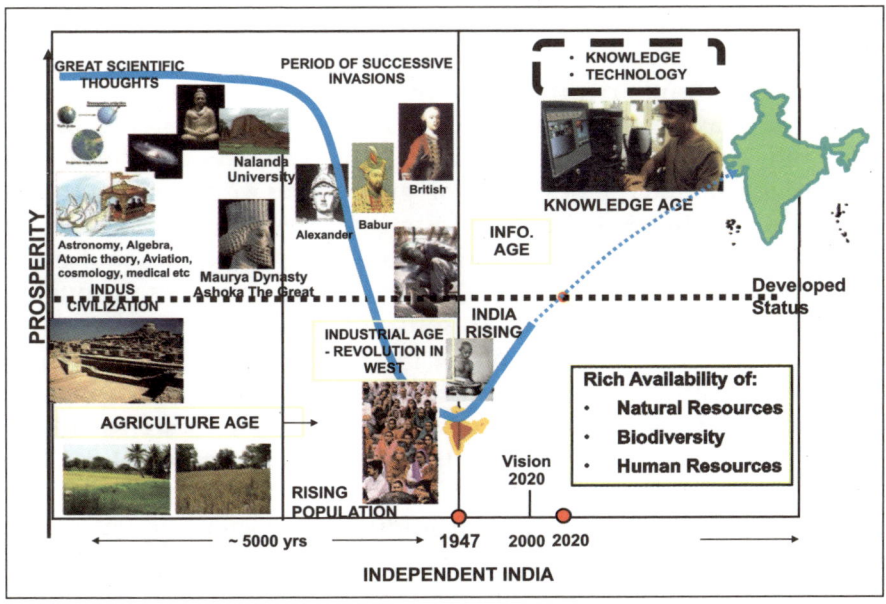

Fig. 1.1: India's Prosperity Dynamics

freedom movement was driven by patriotism and sacrifice with unity of minds. There was an urge in many Indians to excel the foreigners in every field of life. This led to many Indians showing their talent not only for freedom but also in scientific discoveries. Prof. Chandrasekar, the Nobel Laureate says in his biography, 'Chandra' by Kameshwar Wali: "Before 1910, there were no Indian scientists of international repute, but after the First World War, between 1920 and 1925, suddenly, five scientists of international reputation (JC Bose, CV Raman, Meghnath Saha, Srinivasan Ramanujan, and Prof. Chandrasekar himself) emerged. I myself have associated this remarkable phenomenon with the need for self-expression, which became a dominant motive among the young during the national movement. It was part of the national movement to assert oneself. We could show to the West in its own realm that we were equal to them."

The first vision of the freedom movement generated selfless great leaders in every walk of life because only the 'Nation' was there in their heart.

1.2.1 Hurdles in the Path of Development

Brain Drain and Resource Drain

As the country was in the clutches of the foreign rulers, India could not participate and flourish at the time of industrial revolution whereas the West capitalized this opportunity. The West became competent in producing large scale consumable products. The two world wars also gave impetus to these countries to produce new technological military products. Those industries, in the developed countries, have constantly created high technology systems during the cold war to establish a high level of competence. Developed countries retained the best technology products with themselves and offered second-rate technology systems to developing countries to maintain the gap. They also introduced control regimes and trade barriers to weaken the progress of developing countries producing high technology products. The resources of the developing countries got drained due to the import of moderate systems to meet their immediate requirements. Having lost valuable resources to import of moderate technology systems, developing countries were unable to invest in high-tech research and the gap widened (Figure 1.2). The situation persists even now.

Moreover, the highly talented graduates and scientific manpower in the developing countries migrate to developed countries that provide adequate opportunities for growth. These migrated manpower help to develop new technologies and systems for the

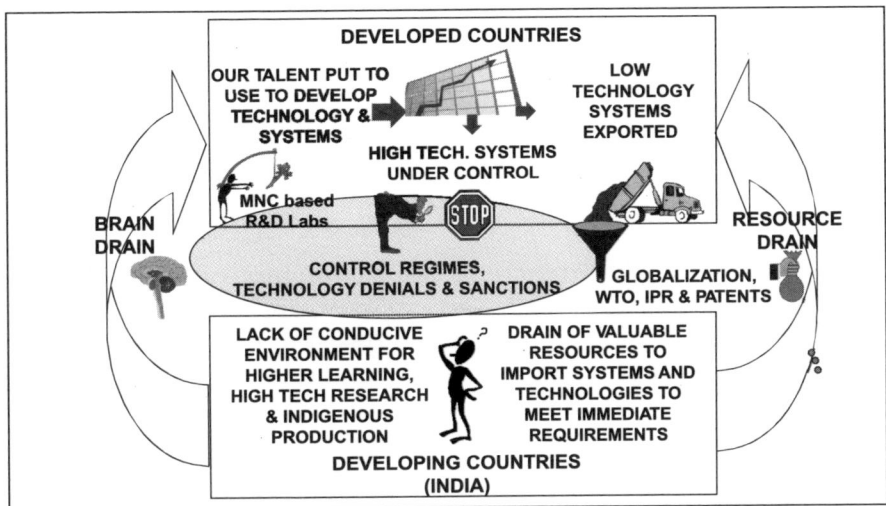

Fig. 1.2: Hurdles in the Path of Development

developed countries to keep their competitive edge. Moreover, the developed countries introduced control regimes and technology denials to ensure that such advanced technologies developed through migrated manpower do not get passed on to the developing countries. This means that the developed countries keep an edge on technology, industrial production with higher market share and economic growth continually, and the developing countries lose their skilled manpower, industrial production and resources, and continue to be economically behind them. This is a major challenge to a developing country that aspires to become a developed country. Indian youth need to understand the trap.

Security concerns

Many of the advanced countries, by virtue of their large scale industry and technology strength, emerged as strategic powers. Post independence, India put enormous effort to make itself a thriving technology hub. It is also pertinent to note that India is a multicultural and multilingual society with a vast population. India's unique stature as the largest democracy in the world desiring universal peace and co-existence and its philosophy of Ahimsa Dharma (Non Violence) are well known to the world. Still, India's relationship with neighbouring countries is constantly disturbed by some issues or the other, disturbing the peace. The threat that India perceives has a very wide spectrum. The world is more concerned with the growing security threats. Security concerns are of two types -

global security and internal security. Global security concerns are nuclear arsenal and carriers, chemical and bio-terrorism, cyber crime, border disputes, etc. Internal security issues are the emergence of religious fundamentalism, ethnic conflicts, Maoism, naxalism, state sponsored terrorism, etc.

1.2.2 Changing Situation

India converted these threat perceptions into motivation to develop high technologies leading to self-reliance and security. Indian scientists living in India demonstrated that they can develop critical technologies through indigenous effort and prove to those countries who denied the technology that "We can do it". Therefore, technology denial and control regimes had little impact on programmes of national importance such as Space, Nuclear and Missile Programmes. India today is a nuclear weapon state, self reliant in space technology and missile power.

The country is experiencing a dynamic situation in economic growth. There is a slowdown which needs to be corrected through a resolved solution by united effort. Now, talented human resource is returning to India after acquiring knowledge from all over the world because today's emergent India offers a whale of opportunities in hi-tech research and development and industrial growth. India is also producing large skilled manpower which is also available to the world. With appropriate government policy for science and technology, networked industry, academy and R&D consortiums, rationalised tax and duties and investment for industrial growth and infrastructure, aggressive market both inside and outside India, trade agreements in the market zones such as ASEAN, Africa and South American countries, the economy can be revived to commensurate with our Vision 2020.

1.3 TECHNOLOGY VISION 2020

The movement of independent India that started in 1857 brought unity of minds, focused action, strong visible leadership, mass movement driven by patriotism, and perseverance. It also gave a great message to the world that a nation can achieve independence through non-violence. The freedom movement was the first vision of India. The need of the hour is to make India a developed nation and to regain the past glory. This becomes our second vision for the nation. The objectives of the second vision were to be self-sufficient in

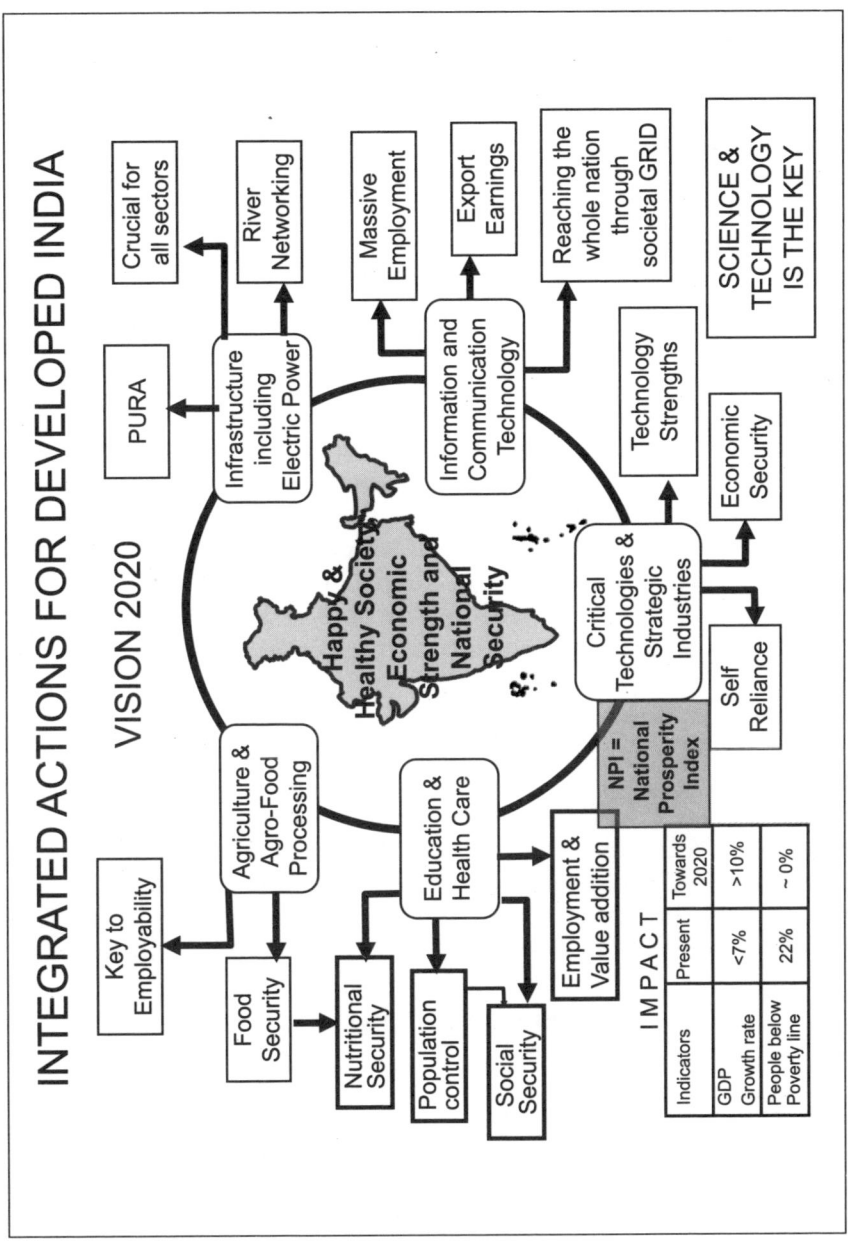

Fig. 1.3: Integrated Actions for Developed India

food production, achieve sustained GDP growth rate above 10 per cent, poverty below 10 per cent and a literacy of 100 per cent.

In order to achieve the above vision of a developed nation, TIFAC (Technology Information, Forecasting and Assessment Centre) of Department of Science and Technology undertook the study on Technology Vision 2020 to identify the specific areas for development and road map to achieve the desired results. While detailed action plans arrived at in 17 volumes, five key result areas were identified based on India's core competence as shown in Figure 1.3. These areas are: (i) Agriculture and Food Processing for food security; (ii) Education and Healthcare for social security; (iii) Information and Communication Technology and mass employment; (iv) Infrastructure including Electric Power, networking of rivers and PURA for economic development; and (v) Critical Technologies and Strategic Industries for self reliance. These sectors are closely interlinked and can together lead to strengthening of the economy and national security. Integrated actions in the above areas, if implemented, will result in GDP growth rate of more than 10 per cent from the present level sustained in the same range for at least 10 years, and the number of people below the poverty line to less than 10 per cent and increased literacy and employment.

India began its journey of economic progress by concentrating on non-strategic sectors like agriculture, dairy, pharmaceuticals, telecommunications, power generation and cars because these were the dire needs at one point of time. With continuous efforts, India became near self-reliant in these non-strategic sectors. Development of these sectors paved the way for the ultimate growth of connected sectors comprising infrastructure development, strategic sectors and other high technology areas. The road map outlined in the Technology Vision 2020, approved by the Government of India in 1996 has some impact in implementation. A vigorous action plan will have a great impact to sustained growth of economy. Instead, the Government funds are being utilised in popular schemes which will have only a temporary effect.

1.4 ECONOMIC GROWTH

In an increasingly interdependent world, all countries are pursuing policies to work together to achieve mutual benefits. Rapid technological changes, while making transactions more seamless, will reinforce the process of global integration. India is the largest democracy in the world and has emerged as a strategic power in the international arena and is making rapid strides in the triad of Space, Defence and Nuclear energy. India has emerged as a leading hub for multiple sectors such as Information and Communication Technology, Automobile Engineering and Pharmaceuticals.

The reform process in India was initiated in the year 1991 with an aim to accelerate the swiftness of economic growth and the eradication of poverty. Government enforced a planned switch-over to a more open economy with greater reliance upon market forces, larger roles for the private sector including foreign investment, and restructuring the Government's role in the economy. An important feature of India's reform programme was that it laid more stress on gradualism and evolutionary changeover rather than rapid streamlining. The economic reforms of 1991 introduced far-reaching measures which changed the pattern of the economy. These changes were related to (a) the dominance of the public sector in the industrial activity, (b) discretionary controls on industrial investment and capacity expansion, (c) trade & exchange controls, (d) limited access to foreign investments and public ownership and (e) regulation of the financial sector. The reforms unlocked India's enormous growth potential and unleashed powerful entrepreneurial forces.

After the economic revival in 1991, higher growth rate has been achieved and sustained for many years in manufacturing and services. Commissions have been set up for manufacturing, knowledge, agriculture to stimulate further growth taking into account the increase in population and skill requirement for the youth. The education bill was introduced for compulsory education of the children up to the age of fourteen years. The GDP growth rate of over 9 per cent put the economy in the right path. According to the estimates by Goldman Sachs, India would be among the top three largest economies of the world in 2050 (Figure 1.4). However, we see a decline in the last two years of the GDP Growth rate going from 9 per cent to just below 6 per cent. Industrial output has declined. Fiscal deficit is to be brought down.

Analysis by the world forums started giving a negative picture of the economical situation of India. What is being experienced today may be a temporary phase and India is poised to become an economically strong nation. Growth oriented government policies, united efforts of the youth to build the nation through major projects using technology as a tool for development, rural development through PURA and building knowledge society will resolve the problems being faced by the nation.

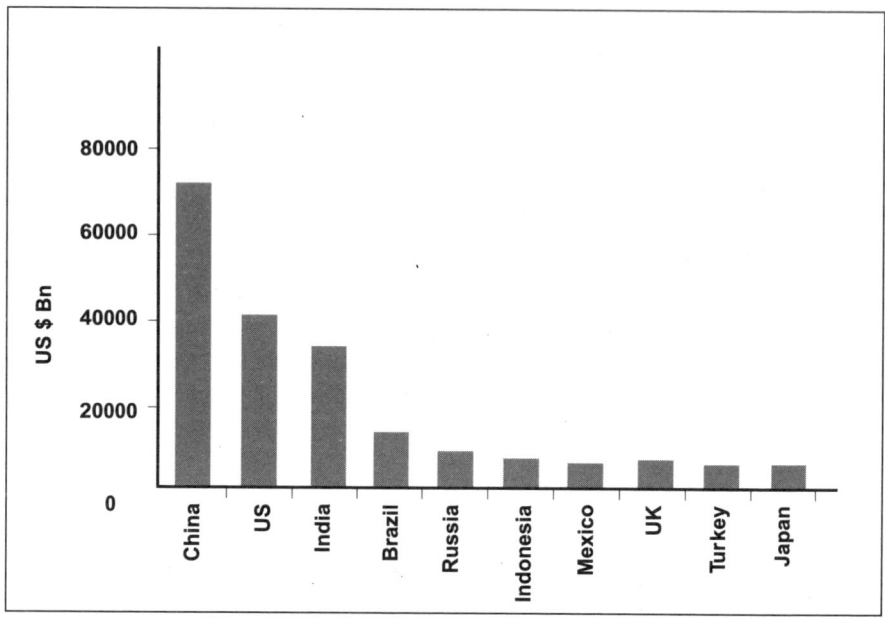

Fig. 1.4: GDP Growth Rate Predictions & Status in 2050

1.5 KNOWLEDGE POTENTIAL

India has a vast resource of knowledge and consequently has generated skilled professionals in every field. There are over 610 universities and 32,000 colleges which are continuously training professionals (Figure 1.5). Talented youth is approximately 580 million and they contribute to the strength of this country. Another advantage that India has is that it has the second largest English-speaking population in the world, hence, it provides a platform for Indian professionals to perform in industries having a multinational environment. The world is moving towards knowledge-based products and services and India will make apt use of its strength in knowledge.

In the societal transformation, the world has witnessed four societies – Agriculture, Industrial, Information and Knowledge. Each society contributes to economy. India is blessed with large skilled human resource which has become our core competence. As innovative knowledge products and services are dominating the society, the knowledge core competence of India becomes a blessing to the economy. Thus, in today's world, technology and knowledge workers are the key resource for enhancing the economic

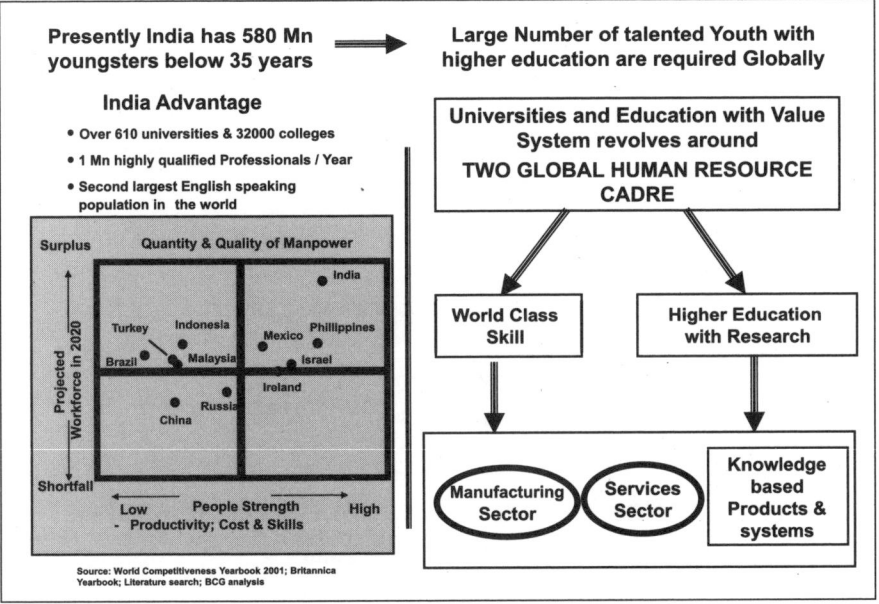

Fig. 1.5: India's Knowledge Potential

development and the future growth. This needs a vision for science and technology and a policy framework for education, research and a scientific tempo in the country.

1.6 SELF-RELIANCE IN SCIENCE AND TECHNOLOGY

The need of the hour is to have a vision and a strong partnership between R&D institutions, academia and industry that will give the necessary scientific and technological strength to the nation (Figure 1.6). Education system needs a large scale revision to generate creativity, value system and patriotism. The active participation of academic institutions in scientific and high technology research will give impetus for higher level of knowledge and will reverse brain drain. Participation of industry will lead to improved productivity and higher GDP. In the globalised world, it is essential to make high technology products with a competitive edge to get market share. Academic institutions should become technology incubators for industries with strong synergy with R&D laboratories. The effect of this synergy among the three players with appropriate government policies will lead to excellence in science and technology and thereby in economic prosperity and military might.

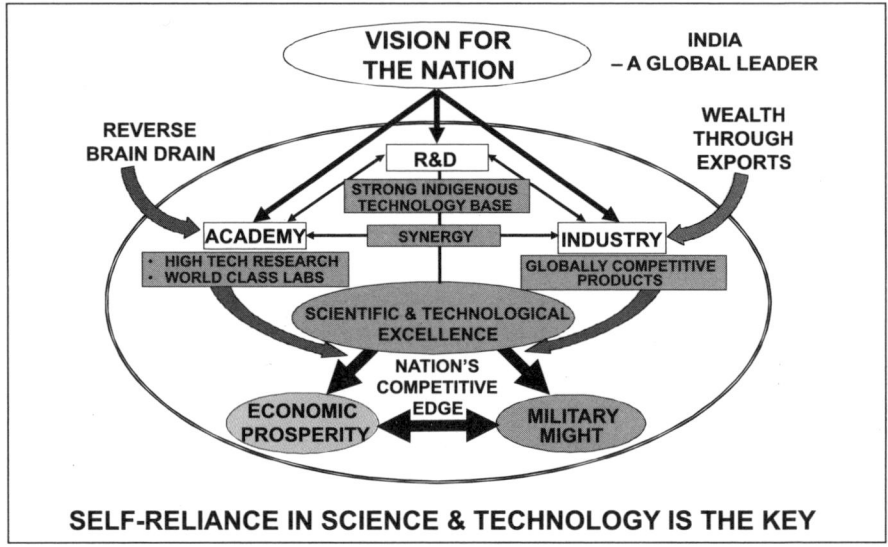

Fig: 1.6: Need of the Hour

1.7 CONCLUSION

We belong to a great gene, which nurtured science and technology, unique culture and value system rooted in the world's glorious past. We have to activate our DNA to perform the best to make India a global leader. We have made advancements in many fields to become a player in the international world. Many more things have to be done. The Government introduced commissions for agriculture, knowledge and manufacturing which constitute the economic growth of the nation. But the impact is yet to be seen due to divergent thoughts and actions. We need to unite our thought process, keeping in mind that the Nation is bigger than individual.

A strong economy and strategic power are the two most important components for any country to make its presence felt all over the world. Both these factors are dependent on the state of technology. Strong economies are derived through infrastructure, industrial growth and agriculture that are solely dependent on the use of new technologies. Strategic power is derived through high-technology systems which are in place. Special emphasis must be given to thrust areas like agriculture and food processing, education and employment, healthcare, infrastructure including power, networking of rivers, rural development using PURA, and development of strategic industries for self-reliance. These actions will provide food, water, clean environment, social, health and economic securities to India. From this platform, India is poised to become a global leader.

REFERENCES

1.   Great Indian Acharyas, Rishi - Scientists of India, Indian Institute of Scientific Heritage, Trivandrum, www.iish.com.
2.   Great Indians who did Great Jobs, Stephen Knapp, http://www.stephen-knapp.com/great_indians_who_did_great_things.htm
3.   Grand Anaicut, http://en.wikipedia.org
4.   Bodhidharma From Myth to Reality, Joseph Aranha, International Seminar on the Contributions of Tamils to the Composite Culture of Asia, 16 – 18 January 2011.
5.   India Revisited, Goldman Sachs Asset Management, White Paper June 2010.
6.   Inclusive and qualitative expansion of higher education, 12th Five Year Plan, University Grants Commission.
7.   English speakers (most recent) by country
     http://www.nationmaster.com/graph/lan_eng_spe-language-english-speakers
8.   World Competitiveness Year book 2001, Britannica year book, BCG Analysis.

PART 2

Mission Mode Programmes And Technology Push

After independence, the Indian Government under the visionary leadership of Pandit Jawaharlal Nehru, started its development through the Five Year Plans. The first five year plan (1951-1956) aimed at self sufficiency in food by concentrating on agrarian sector through infrastructure investment in dams, irrigation and power. With the result of the implementation of the first five year plan, the achieved growth rate was 3.6 per cent annual gross domestic product as against the target growth rate of 2.1 per cent. The net domestic product went up by 15 per cent. The second five year plan (1956-1961) focussed on industry development including a number of public sector undertakings and assumed a closed economy. Hydro electric power and steel plants at Bhilai, Durgapur and Rourkela were established. Coal production was increased. Large scale defence PSUs established. In the Third five year plan (1961-1966), stress was on agriculture and improved production of grains and cereals. In the fourth five year plan (1969-1974), the Government nationalised fourteen banks and started the green revolution and industry development. Fifth five year plan (1974-1979) laid emphasis on employment and poverty alleviation and increased industrial productivity. The Government entered into power generation and established national high way road systems.

2.1 SCIENCE AND TECHNOLOGY IN INDIA

India had started its mission in early 60s to achieve self reliance in the core areas. The Green Revolution introduced many technologies for higher productivity of grains and cereals leading to 241.6 million tonnes. In power generation India achieved 207,006 MW through thermal, hydel, wind and nuclear power stations. Development of supercomputers such as PACE++ by DRDO and PARAM by C-DAC placed India in the Supercomputer world. Revolution in IT and communication with 943 million connections helped India in the path of economic recovery utilising large skilled manpower. The space and defence programmes have achieved spectacular milestones leading to self-reliance and development of critical technologies. The future of India lies in the growth of science and technology, especially in the areas of nanoscience, molecular biology, environmental science, pharmaceuticals and material science. The growth of science and technology in India is brought out in Figure 2.1.

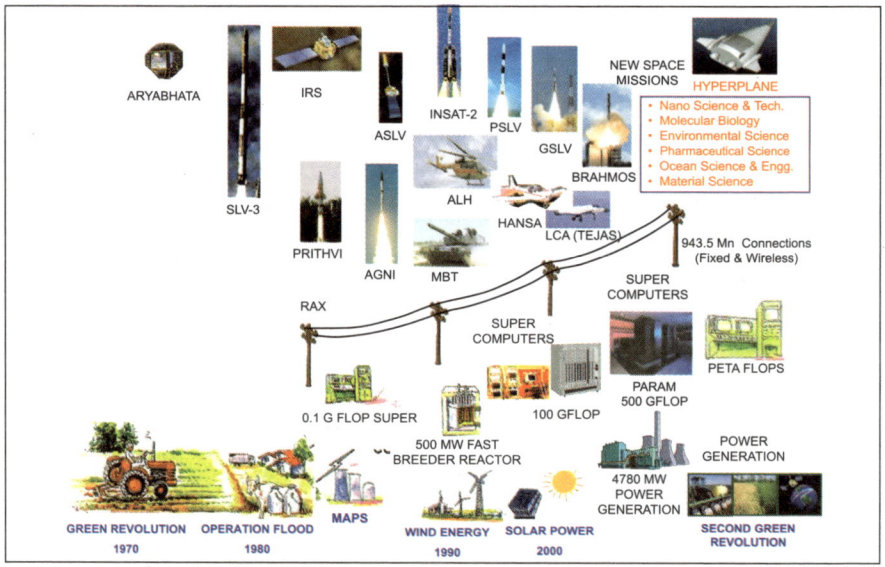

Fig. 2.1: Science and Technology in India

In recent times, we have seen how some visionary leaders emerged by leading successful mission mode programmes for India. The succeeding paragraphs enumerate few of such important mission mode programmes.

2.2 AGRICULTURE – GREEN REVOLUTION

After independence and till the early 1960s, India lived on a ship-to-mouth condition in which food grains were imported from USA under PL 480 Scheme. After the haunting memories of famine and heavy shortage of food in the country, the First Green Revolution emanated during the 1970s from the political leadership of C. Subramaniam, the then Food Minister. The team led by the renowned Nobel Laureate Dr. Norman Borlaug, Dr. MS Swaminathan and the agricultural scientists used their skills in genetic engineering, and with the help of farmers produced food grains to meet the need.

Through the first Green Revolution, there were dramatic changes within a span of ten years. With the help of technology, new and better crops were generated and consortiums were formed where farmers, agriculture scientists, and village authorities were involved. Through an effort of historical magnitude, India attained near self-sufficiency in food through "Seed to Grain" mission. As part of this first green revolution, the country has now been able to produce over 241.6 million tonnes of food grains per year now.

The political leadership along with scientific leadership has been able to build the capacity of our scientists, researchers and farmers to take up the mission of "second green revolution". This is indeed a knowledge graduation from characterization of soil to the matching of seed with the composition of the fertilizer, water management and evolving of pre-harvesting techniques for such conditions. The domain of a farmer's work would enlarge from grain production to food processing and marketing. The technological steps taken in first and second green revolutions are brought out in Figure 2.2. By 2020, India would be required to produce 400 million tonnes to cater to the population at that time. The increase in the production would

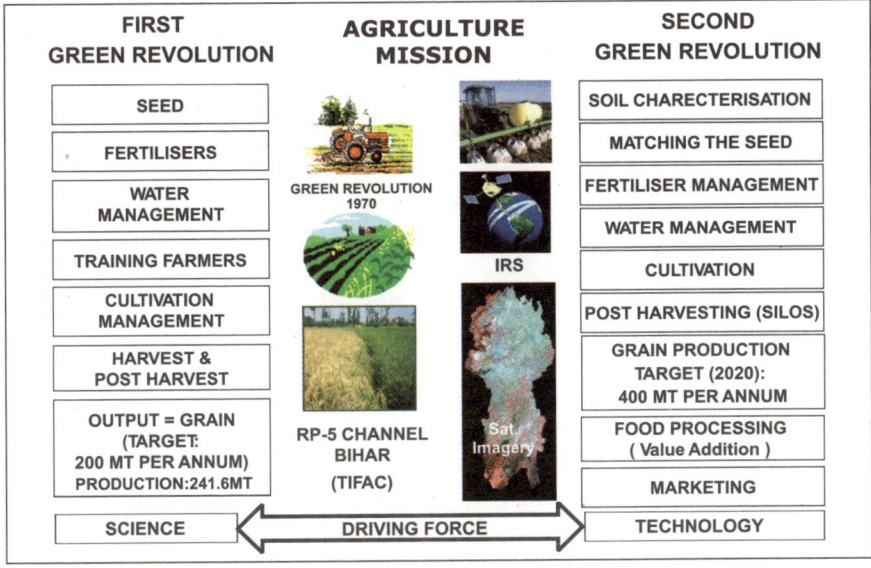

Fig. 2.2: Green Revolution

surmount many impeding factors such as shortage of water, reduced availability of land due to urbanisation and also reduced agricultural workforce. Our agricultural scientists and technologists in partnership with the farmers have to work for increasing the average productivity per hectare by three times compared to present productivity. This was experimented successfully at RP 5 Channel Bihar under the scheme of TIFAC. The type of technologies needed would be in the areas of development of seeds that would ensure high yield varieties even under constraint of water, introducing multiple crops, fertilizer management, post harvesting and storage techniques with reduced waste, food processing, marketing the yields appropriately to the farmers' gain and effective distribution system.

2.3 MILK – OPERATION FLOOD

Varghese Kurien

As a young engineer, Dr. Varghese Kurien was appointed in Government Creamery unit at Anand, in the year 1949. Though the demand was not much but still the Government wanted to make powder out of Buffalo milk. The entire world is making powder out of cow milk and there was a feeling that skimmed milk and milk powder can be made only from cow milk. The challenge was to convert buffalo milk into powder. India had more buffalos and fewer cows. Kurien with his team succeeded in processing buffalo milk to make milk powder and condensed milk. This was a major scientific and technological breakthrough. Prior to this, in 1946, Mr. Tribhuvandas Patel formed five village cooperatives under the Kaira District Cooperative Milk Producers Union Limited (KDCMPUL). Dr. Kurien gave up his job and on the persuasion of Mr. Tribhuvandas, joined hands with him to establish an advanced dairy plant within a span of eleven months. It was owned by farmers and was dedicated to the nation by Prime Minister Jawaharlal Nehru in 1955. Thus the cooperative movement started. In 1957, the farmers registered the trade mark "AMUL" and their brand 'Amul' became a household name.

Milk was scarce in urban areas but abundant in rural areas. Milk being a perishable item, its collection, transportation and distribution in urban areas was a problem. Middlemen, misinformation and lack of infrastructure existed and afflicted the milk distribution process. The White Revolution in the form of Amul dairy cooperative movement showed the way to resolve these problems.

These cooperatives were autonomous, transparent, democratic and had a commitment to quality and honesty. Each member was being paid on the basis of the fat content of the milk brought by them. It brought a revolutionary change in the whole milk processing industry from the hybrid milk cattle, the artificial insemination, the animal health system to the scientific manufacture of the animal feeds. It was the best in the country at that time and opened up avenues for other interested states. In 1965, The National Dairy Development Board (NDDB) was created to extend the success story of the cooperative movement in other states of India and Dr. Kurien was made the Chairman of NDDB.

Today, India is the largest producer of milk in the world with a production of 122 million tonnes and also has a largest cattle population (75 mn adult female cattle & 55 mn adult female buffalo). The success of mission mode programmes was due to the apt use of technology available at that time and pooling of resources for a common cause through the cooperative movement. The scientific and professional management established a direct linkage between the producer and the consumer. The future of the Indian dairy industry will further brighten with long-term export plans. In order to remain globally competitive, India must maintain the current growth rate of milk production. The availability of milk and milk products in abundance is a success story. Dr. Varghese Kurien, the Father of White Revolution in India, led this cooperative movement and dairy development industry in the country.

2.4 INDUSTRY INITIATIVE

The first visionary industrialist in pre-independent India was Jamshetji Nusserwanji Tata. He envisioned India as a manufacturer of steel. He wanted to establish a steel industry in Bihar where raw materials were available in abundance. The British were taking this raw material to Great Britain for converting it into steel. Tata thought that if he could get the technology for converting raw material to steel in India itself, the country will benefit

Jamshetji Nusserwanji Tata

through industrial development. He needed the help of the British who were his best friends. He boarded a ship to London to meet his English friends whom he thought could help him in providing the technology and expertise. It was September 1893, Swami Vivekananda was also travelling in the same ship, to Chicago to attend the World Parliament of Religions. Nusserwanji Tata explained his intention to Swami Vivekananda, who in turn told Nusserwanji Tata "When you are trying to make a steel industry, think of building an institution which will teach materials science". Tata accepted the suggestion. After reaching London, he attended a dinner hosted by his friends from whom he expected help. During the dinner, Nusserwanji requested his friends to help him in establishing a steel industry with their technology. There was much laughter. They said "if India can make steel, then the British will eat it". Offended by their statement and denial of steel technology, he walked out in anger. From London, he immediately proceeded to USA to meet other friends, who could help him. His struggle started giving fruits. While the steel industry was being built, Nusserwanji remembered Swami Vivekananda's words to establish an Institute for research in material science. He then built Indian Institute of Science (IISc) at Bengaluru with a focus on Science and Engineering including metallurgy and material science. The steel industry in Jamshedpur and IISc in Bengaluru came out of his vision. India became a country of steel industry and IISc a world-class institution. Today, the Tata group of industries has purchased the steel industry CORUS from the same UK, which refused the technology long ago. In United Kingdom itself, most of the steel industries are owned by Indians, one of them, Lakshmi Mittal, is the richest man of UK. Steps taken by Nusserwanji Tata paved the way for industrial growth in the country in many fields.

Another great visionary who occupies a distinction for industrial development in India is Walchand Hirachand. In the pre-independent India, Walchand thought about a large scale industrial build up in multiple products and services to generate talent and infrastructure to compete with the West. He established Hindustan Aircraft Company to produce Aircraft which was later renamed as Hindustan Aeronautics Ltd (HAL). He pioneered the Hindustan Construction Company,

Walchand Hirachand

Hindustan Shipyard, Automobile Industry, Sugar Industries, Industry for machinery and many more to revolutionize industrialization in India. He undertook endeavours with persistence over scepticism and hope over despair. He sowed the seeds for the growth of large industries, employment to many, and brought a culture to support the Indian population.

In Walchand's words: "I desire to see Indian ships carrying India's maritime trades to the four quarters of the globe. I am anxious to see that these ships are controlled by Indians as well as managed by

Indians and fully manned by my own countrymen. I yearn to see the Indian flag only as a matter of right, in my own homewaters. I long for the day when I can take my own countrymen in super-Indian "Victorian" and in super-Indian-"Normandies" to all parts of the world".

2.5 NUCLEAR ENERGY

In the words of Homi Jehangir Bhabha "The acquisition by the man of knowledge of how to release and use atomic energy must be recognised as the third epoch of human history". In less than a fortnight after independence, on 26 August 1947, Pandit Jawaharlal Nehru set up the Board of Research on Atomic Energy with Homi Bhabha as Chairman. Following the Atomic Energy Act passed on 10 August 1948, the Atomic Energy Commission was set up. Bhabha envisioned nuclear energy for power generation in a span of two decades and therefore established an institute to train human resource specialising in Nuclear energy and associated industrial capabilities. His tireless efforts, and subsequently that of his successors, made the country self-reliant in nuclear energy. Fast breeder reactor, fuel processing capabilities, pressurized heavy water reactors, etc., are remarkable developments that testify to the advances in nuclear technology. The Fast Breeder Reactor (FBR) is a fast neutron reactor designed to breed fuel by producing more fissile material than it consumes. India has an active development programme featuring fast thermal breeder reactors. India's first 40 MW Fast Breeder Test Reactor (FBTR) attained criticality on 18 October 1985. Thus, India became the sixth nation to have the technology to build and operate after USA, UK, France, China and Russia. India has developed the technology to produce the plutonium-rich U-Pu mixed carbide fuel. This can be used in fast breeder reactor.

The first peaceful nuclear experiment was conducted at Pokhran on 18th May 1974 when Indira Gandhi was the Prime Minister, a bold initiative to make India a strong nation. The tremendous support given by the Government led to the establishment of many reactors for power generation and also to achieve self-reliance in critical technologies. The country had seen spectacular progress by focussed mission projects by the Department of Atomic Energy. This led to the preparations for the second series of five tests, scheduled to be carried out in 1995-96. DAE and DRDO worked together to prepare the tests and we were ready. But the tests could not be done due to international situation leading the Government to backfoot. Hence, in May 1996, the outgoing Prime Minister Narasimha Rao told Atal Bihari

Figure 2.3: Nuclear Programme

Vajpayee who was the Prime Minister designate about the status and advised him to conduct the nuclear experiments. The two leaders put the Nation above political considerations and wanted to make the country strong. Unfortunately, the tests again could not be conducted as planned, as Vajpayee's Government survived only for thirteen days in May 1996. Later, in March 1998, when Vajpayee became the Prime Minister for the second time, he quickly ordered the resumption of preparations for underground tests at Pokhran. The preparations were carried out under camouflaged net so that foreign satellites were not able to know what was happening. The second peaceful nuclear experiments (a series of five tests called Shakti) were conducted on 11 and 13 May 1998. These include thermo nuclear device of 40 kt, fission device of 15 kt, experimental boosted fission device of 0.3 kt, experimental device of 0.5 kt and experimental device U233 of 0.2 kt. Prime Minister Vajpayee made an announcement about the successful conduct of the underground nuclear tests in Pokhran range declaring it as a moment of pride. Prime Ministers Indira Gandhi, PV Narasimha Rao and Atal Bihari Vajpayee were responsible for India becoming a 'Nuclear Weapon State'. The nuclear policy is based on the principle of 'No first use' of nuclear weapons and not to use them against non-nuclear countries, and also to have only minimum credible deterrent. The enunciated nuclear policy expresses India's firmness in propagating peace as a responsible nation and at the same time making the country a strong nation – Strength respects Strength.

The R&D efforts in nuclear science helped India to establish its first nuclear reactor to generate power. Today, India generates 4,780 MW electric power. With continuous developmental efforts and a nuclear agreement with IAEA and Nuclear Supplier Group (NSG) of countries, India is targeting to produce 40,000 MW electric power by 2020. Power reactors of today mostly use a fissile fuel called uranium-235 (U-235), whose 'fission' releases energy and some 'spare' neutrons that maintain the chain reaction. But only seven out of 1,000 atoms of naturally occurring uranium are of this type. The rest are 'fertile', meaning they cannot fission but can be converted into fissionable plutonium by neutrons released by U-235. India has the capability to use thorium cycle based processes to extract nuclear fuel. This is of special significance to the Indian nuclear power generation strategy as India has more than one-third of the world's reserve of Thorium on the beaches of Kerala and Orissa that can fuel nuclear reactors for an estimated 2,500 years. The byproducts of the liquid-based thorium reactor are desalination agents for potable water, hydrogen, automotive fuel cell, etc.

Fig. 2.4: India's Space Profile (Source ISRO)

2.6 SPACE VENTURE

Dr. Vikram Ambalal Sarabhai pioneered India's Space Programme. Bhabha supported Sarabhai in setting up the first rocket launching station at Thumba, primarily because the magnetic equator of earth passes through this place. This development furthered the indigenous capability for satellite launching from low-orbiting to synchronous orbit. The first major mission mode programme was thedevelopment of SLV-3 including the development of launch vehicle and staging technologies. All the efforts of ISRO were concentrated on achieving this mission of successful launching of SLV-3. Prof. Satish Dhawan, Dr. Brahmprakash and many senior scientists, staff, technicians worked round the clock for developing and proving the 44 subsystems which constituted SLV-3. The first launch on 10th August 1979 did not succeed due to the failure of second stage control system. Though it was a setback, the scientific community of ISRO learnt very important lessons on technology, quality management and courage to overcome failures. The second launch of SLV-3 on 18 July 1980 was successful in injecting the Rohini satellite in low earth orbit and made India attain indigenous launch capability and become an exclusive member of the Space Club.

Augmented Satellite Launch Vehicle (ASLV) enhanced the capability to inject larger satellites for scientific experiments. The major breakthrough in launch vehicle technology came with the launch of Polar Satellite Launch Vehicle (PSLV) with the capability of launching multiple satellites in a single mission. Launch of Geosynchronous Satellite Launch Vehicle (GSLV) has established a unique position for India to orbit INSAT class of missions in a geostationary orbit. Indian Remote Sensing satellite (IRS) series provide high-resolution images which are required for the management of natural resources. The Indian National Satellite (INSAT) series meets the communication, TV and meteorology demands of the country. With PSLV operational and GSLV coming on line, ISRO has established not only self-reliance with launch and satellite capability but can also offer cost-effective launch service to different countries.

Dr. Vikram Sarabhai, Prof. Satish Dhawan and their able successors, with their vision and meticulous planning of programmes, led the nation towards self-reliance in space technology in satellites, launch vehicles, mission management and utilization of space for the benefit of the countrymen. The mission of Chandrayaan-I, in the first attempt itself, was an astounding success. Mini SAR camera onboard had taken a number of photos of the moon that found frozen ice, for the first time, on the surface of the moon opening up further research. The recent launch of radar imaging satellite RISAT-1 with synthetic aperture radar payload is another major milestone for all weather surveillance. The success of ISRO programmes in launch vehicles, satellites and implementation of applications in remote sensing and communication can be attributed to strong vision, technological leadership and mission mode management.

2.7 MISSILE PROJECTS

Integrated Guided Missile Development Programme (IGMDP) with an aim of developing five different missiles – Prithvi and Agni for strategic applications, Trishul and Akash for air defence, and Nag as anti tank missile, started in July 1983 with DRDL as the prime development laboratory. This was a major mission to indigenously develop the most advanced technologies in propulsion, structure, composites, guidance, control and warheads. Many new unconventional concepts were tried out for the first time to give a leap-frog in technology and performance. Among many new technologies established during this programme, the development of re-entry structure, guidance and control for Agni gave long range strategic missiles in different versions from 700 to 5000 km. Prithvi and Agni have been inducted in the strategic command. The surface to air missile, Akash with a multi-

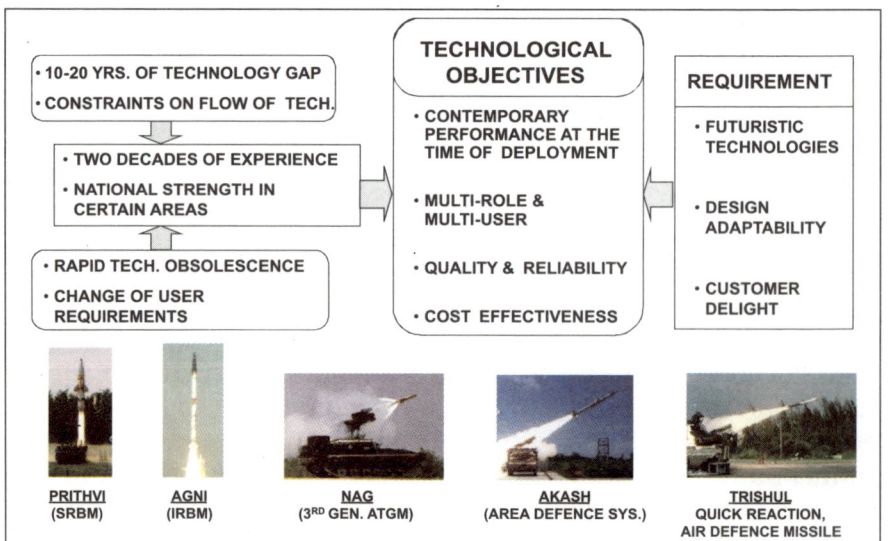

Fig. 2.5: Evolution of IGMDP

target handling capability is in production for the Air Force as area defence weapon. The third generation anti tank missile Nag has completed its development with advanced imaging infrared guidance for fire and forget capability with top attack trajectory.

Through the IGMDP, India has mastered several critical missile technologies which were denied under the MTCR. Unique concepts used in the design to give a leap frog in performance, contemporary to missiles of advanced countries, proved the capability of Indian minds to excel during difficult times. The strong leadership, mission mode approach and effective programme management integrating the stakeholders resulted in successful results. The details of technologies developed indigenously and the methods used to develop are discussed under the later under the section of Missile Technology.

Another mission mode programme was BrahMos. This Indo-Russian joint venture Supersonic Cruise Missile which started in 1998 harnessed the strengths of Indian missile technologies with that of Russian Institutes and proved to the world that a joint venture of advanced technology can lead to a high performance product in the shortest possible time with a far-reaching capability. No other country has universal and unique supersonic cruise missile like BRAHMOS. Thus, India became a country to possess technology better than that of the developed countries through this joint venture. BRAHMOS is a 290-km range supersonic cruise missile flying at a speed of 2.8 Mach. The missile belongs to fire-and-forget class and can be fired from multiple platforms against multiple targets using varied cruise trajectories. It has been successfully test-fired from various platforms like ships and mobile autonomous launchers. The missile has been inducted in the Indian Armed Forces. Indian Army is the first land force in the world to possess a supersonic cruise missile regiment.

Another mission mode programme was the development of a two layer ballistic missile defence in exo atmosphere and endo atmosphere. Successful trials of interception proved the missile system, and the radar and control centres employed for integrated action and quick engagement. The system with high level of sophistication with long range radars, complex mission calculations, quick successive launch of interceptors, continuous tracking and control of both target and interceptors at exo and endo atmosphere regions, has been demonstrated for its effectiveness. The system is ready for induction.

India today is a missile power with two types of nuclear tipped strategic missiles covering up to 5000 km and the most advanced precision strike supersonic cruise missile BRAHMOS against sea and land targets, already inducted in the services. A credible two layer ballistic missile defence system has been tested successfully to defend against any incoming nuclear missiles. India is now self-reliant in design, development and production of any type of missile and is also technologically competent.

2.8 INFORMATION AND COMMUNICATION TECHNOLOGY

2.8.1 Information Technology

The Indian IT sector continues to be one of the most promising sectors of the Indian economy showing rapid growth and potential. According to a report 'Perspective 2020: Transform Business, Transform India' prepared by McKinsey for NASSCOM, the exports component of the Indian industry is expected to reach US$ 175 billion in revenue by 2020. The domestic component will contribute US$ 50 billion in revenue by 2020. Together, the export and domestic markets are likely to bring in US$ 225 billion in revenue, as new opportunities emerge in areas such as public sector and healthcare and as countries including Brazil, Russia, China and Japan opt for greater outsourcing.

IT-BPO sector has become one of the most significant growth catalysts for the Indian economy. In addition to fuelling India's economy, this industry is also positively influencing the lives of its people through an active direct and indirect contribution to various socio-economic parameters such as employment, standard of living and diversity. According to NASSCOM, IT industry has played a significant role in transforming India's image from a slow moving bureaucratic economy to a land of innovative entrepreneurs and a global player in providing world class technology solutions and business services.

The sector is presently estimated to generate $100 billion in FY2012. The export revenues are estimated to have aggregated to US$ 69 billion in FY2012 and contributed 25 per cent as its share in total Indian exports (merchandise plus services), according to a research report 'IT-BPO Sector in India: Strategic Review 2012', published by NASSCOM. The workforce in Indian IT industry will touch 30 million by 2020 and this sunrise industry is expected to continue its mammoth growth.

India is a preferred destination for companies looking to offshore their IT and backoffice functions. Business environment, availability of skilled people and low cost advantage make India a financially attractive location. According to global research group Gartner, India's domestic market for business process outsourcing (BPO) is projected to grow at 17 per cent to touch ~918 billion in FY 2012.

2.8.2 Communication

The Indian telecommunication network is the third largest in the world and the second largest among the emerging economies of Asia. India is among the top ten in the world in the telecommunication network with 943.5 million Subscribers (fixed & wireless). Indian telecommunication has continued to record noteworthy success throughout the year and has emerged as one of the key sectors that have been accountable for resurgent growth of the Indian economy. The rapid growth of the sector has been coupled with proactive policies and decisions taken by the Indian Government, and dynamic involvement of the private sector.

The telecom sector in India is a major contributor to the economy and is a vital employment generating industry for thousands of professionals. With a direct impact on the socio-economic structure of the country, the sector has been able to successfully surpass the targets set up by the policy makers. The dynamism displayed by the government and the private sector for uplifting the telecom sector has been commendable and speaks volumes about the efforts behind the success story.

2.9 PHARMA INDUSTRY

Pharmacology is a science that provides a multilevel approach for treatment starting from the molecular, cellular and tissue levels to complete human or other forms of life. Certain treatments such as neuropathic diabetic ulcers which were considered very difficult to treat so far, have been treated successfully with polymeric skin carrier and skin graft due to the advent of new technologies in pharmaceuticals.

The Indian pharma sector commenced operations with repacking and preparation of formulations from imported bulk drugs but it has now moved on to become a net foreign exchange earner with the capability of producing more than 400 bulk drugs within the country. From a mere processing industry, it has grown into an advanced sector with advanced manufacturing technologies, modern equipment, and stringent quality control in developing products and at an affordable price.

The global production of pharmaceuticals, branded and generics, put together is of the order of $773 billion. India is one of the fastest-growing pharmaceutical markets in the world, and its market size has nearly doubled since 2005. The Indian Pharmaceutical industry turnover has grown to Rs.104209 Crs (US$21.7 billion) and according to the report by Pricewaterhouse Coopers, the sales are expected to reach US$ 74 billion by the year 2020. India has got core competence of producing cost effective Pharma products. It is the third largest market in the world in terms of volume and fourteenth in terms of value. It is time for the pharma industry to envisage a vision for making India number one in the production of drugs. In order to achieve this, it would need to invest more in R&D from the present level of 1–2 per cent of sales, more so in view of the patent regime under WTO from the year 2005. With a surging pharma sector and availability of specialised doctors with the latest state-of-the-art drugs and equipment and quality nursing with care, India is emerging as a healthcare hub for quality medical service at affordable cost.

At this juncture, it is essential that the Indian pharma industry takes a bold step to further extend its partnership with many institutes and academia for quality research in order to have its rightful share in the world market. Also, in the globalisation regime, holding collaboration with a suitable industry partner and leapfrogging the core technologies will provide a further edge.

In a knowledge society, the most important commodity is the human resource empowered with right knowledge. The role of both institutes and industry, is very important for nurturing this talent. Institutes impart valuable training to students and industries provide students the opportunities and suitable platform to perform their capabilities. This combination will take the pharma sector to a new height. The pharma sector coupled with intensive R&D will transform this sector further. The role of research is of paramount importance in pharmacology. Indian chemists, who are highly skilled in activities such as organic synthesis, medicinal chemistry, process chemistry and analytical chemistry are in much demand. The collaboration with Indian Clinical Research Organisations (CRO) would definitely help to cover gaps in capacity, optimization of cost, widened kill base and enhancement of drug development. One of the biggest breakthroughs which science has witnessed in the past decade is the discovery and broad application of gene silencing by RNA interference (RNAi). It has been shown that double-stranded RNA molecules lead to the digestion of homologous messenger RNAs (mRNAs) not only in plants or worms but also in animals and humans. This technology has been widely used as a valuable tool in functional genomics, target discovery and validation.

Its use in these fields is likely to result in better drugs entering clinical trials in the near future.

2.10 CONCLUSION

From the above, we can understand that the mission mode approach with strong leadership, planned strategy, integration of stakeholders, programme management, harnessing the best talents and visible targets lead to successful programmes. After independence, during the last sixty years, India has emerged as a science and technology hub with highly skilled human resources. The growth of high technology will move on exponentially and needs mission mode approach for taking full advantage of technology to establish competitiveness. With Technology Vision 2020 in hand, outlining the road map for developed India, these successful mission mode programmes indicate that India will emerge as one of the foremost technologically advanced nations in the world.

REFERENCES

1. 1ˢᵗ Five Year Plan, http://planningcommission.gov.in
2. Envisioning an Empowered Nation, APJ Abdul Kalam, A Sivathanu Pillai, Tata McGraw Hill, 2004.
3. India 2020 – A Vision for the New Millennium by Prof. APJ Abdul Kalam & YS Rajan.
4. The Green Revolution: History, Impact and Future, H.K. Jain, TX Houston Stadium Press, 2010.
5. Sustainable Agriculture: Towards an Evergreen Revolution, MS Swaminathan, Konark Publishers Pvt Ltd, 1996.
6. An Evergreen Revolution, MS Swaminathan, Crop Sci 46:2293-2303, 2006.
7. India's green revolution, economic gains and political costs, F.R. Frankel, Princeton, New Jersey, 1971.
8. The Green Revolution in India: A Case Study of Technological Change, Parayil Govindan, Technology and Culture, Vol 33, No. 4, 1992.
9. An Unfinished Dreams, Varghese Kurien, Tata McGraw Hill, 1997,
10. Verghese Kurien - The father of white revolution, http://www.spiderkerala.net/resources/8712-Verghese-Kurien-The-father-white-revolution.aspx.
11. The story of Tata Steel, Bombay, Verrier Elwin, Tata Iron and Steel Co.,
12. Walchand Hirachand: Man, His Times and Achievements, GD Khanolkar, Ratanchand Hirachand, 2007.

13. From Fission to Fusion The Story of India's Atomic Energy Programme, MR Srinivasan, Viking Penguin India, 2002.

14. Nuclear Energy Development in India, Anil Kakodkar, 19 May 2008.

15. Indian Space Programme – A Multidimensional Perspective, Edited by K Kasturirangan, Current Science, Vol. 93, No. 12, 25 December 2007.

16. Space Technology for Humanity: A profile for the coming 50 years, K Kasturirangan, ScienceDirect, Vol. 23, 2007.

17. Vikram Sarabhai: A Life, Amrita Shah, Viking Penguin India, 2007.

18. DRDL Dare Devil Days, Prahlada, AV Rangarao, Defence Research & Development Laboratory.

19. IGMDP Integrated Guided Missile Programme, Defence Research & Development Organisation, 2008.

20. Annual Report 2011-2012, Department of Telecommunication.

21. Progress and status of different Telecom development parameters for the month of February 2012, Department of Telecom, No.18-10/2011-STT(T), April 2012.

22. Creative Leadership: Essence of vibrant business, APJ Abdul Kalam, Address to the senior industry leaders and members at the CII-Suresh Neotia Centre of Excellence for Leadership, Kolkata, 17 February 2012.

23. Information Technology, India Brand Equity Foundation, www.ibef.org/industry/informationtechnology.aspx.

24. The IT-BPO Sector in India, NASSCOM STRATEGIC REVIEW 2012.

25. Telecommunication – Brief Introduction, India Brand Equity Foundation, www.ibef.org/industry/telecommunication.aspx.

26. Annual Report 2010-2011, Dept. of Pharmaceuticals, Ministry of Chemicals and Fertilizers, http://pharmaceuticals.gov.in/AnnualReport1011/ch_2.pdf

27. Pharmaceuticals, India Brand Equity Foundation, http://www.ibef.org/industry/pharmaceuticals.aspx

28. Pharma Vision 2020, APJ Abdul Kalam, Address at the Convocation of JSS University, Mysore, 29 June 2010.

PART 3

Futuristic Technologies

*"Inventions and discoveries have emanated
from creative minds that have been constantly working
and imagining the outcome in the mind.
All the forces of the Universe
work for that inspired mind."*

History again and again proved that the ingenuity of individual and intellectual supremacy brought the best to the society's development. Some of the unique inventions and technology revolution have transformed the society and life of human beings. These technological aspects are brought out in this part.

3.1 TECHNOLOGY THROUGH AGES

During the last 5000 years, unique cultures have come into existence as the man continuously attempted to have a better life for himself and the society around him. Particularly in the last two hundred years, the society got transformed from purely agriculture to industrial, and then to information with different manifestations. The Industrial Revolution in the eighteenth century paved the way for economical development making life better with new opportunities and large scale employment. Information technology revolutionised man's life further, forming part of major economy. In addition to the industrially developed countries, the economy also grew in the developing countries due to information revolution. India's prosperity multiplied during the IT revolution taking advantage of large skilled human resource. Recent developments in bio and nanotechnologies and their convergence with information technology opened up greater opportunities for the future. These technological revolutions are presented in Figure 3.1.

Inventions and Inventors

Great scientific inventions during the last four centuries have transformed the quality of our life. So we always remember those inventors. When we see the electric bulb giving light, our thoughts

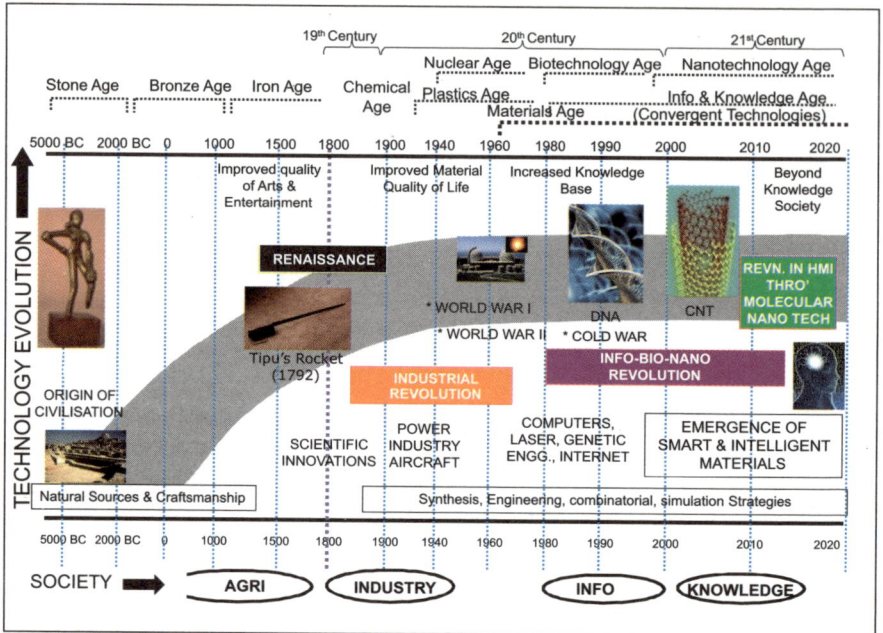

Fig. 3.1: Technology through Ages

instantly go to the inventor Thomas Alva Edison. When we hear the sound of an aeroplane, our thoughts go to Wright brothers who disproved Kelvin's assertive statement, "Heavier than air flying machines are impossible". The telephone reminds us of Alexander Graham Bell.

The Principles of electricity by Benjamin Franklin in 1752, the invention of the first Electric Bulb by Thomas Alva Edison in 1879 and its further developments led to the Sodium Vapour lamp to present day Compact Fluorescent Lamp (CFL) bulbs. The recent development of Light Emitting Diodes (LED) resulted in saving of electricity. Invention of the Radio by Marconi in 1896, first telephone by Graham Bell in 1876 and the first television by John Logie Baird in 1911 had seeded the revolution of telecommunication and entertainment. Today the revolution has touched to mobile handset for wireless communication, High Definition Plasma/LED Television and the mobile TV. The invention of Computer by John Von Neumann in 1946, the concept of World Wide Web (WWW) first proposed by Sir Tim Berners Lee of CERN (The European Organization for Nuclear Research) in 1989 and the revolution in communication with 2G, 3G and 4G spectrum services increased the speed of the Internet, making information handy and available at any point of time. There are other such inventions. The great inventions are depicted on Figure 3.2.

Scientific thoughts and the great scientists of the world brought change, and their names will be remembered forever. Every youth should aspire to be in the galaxy of inventors to be remembered. Scientific research continues in many advanced laboratories worldwide. We attempted to identify unique technologies which will have greater impact on the future. The objective of technological revolution is towards achieving higher level of economic growth, human resource development with employment potential, security of individuals, defence of the country and affordable health care, all leading to prosperity and happiness of the society.

The fundamental scientific disciplines of physics, chemistry and biology are becoming increasingly dependent on each other and the boundaries are expanding. In the last fifty years, there has been a considerable miniaturisation through micro-electronics and recently also through nano-electronics. Integrated applications of laws of physics, chemical properties of materials and biological principles are leading to newer inventions. The inter-disciplinary approach of sciences is leading to evolving convergent technologies. Progress in convergent technologies will bring substantial increase in efficiency and sustainability with reference to Environmental, Social and Economic progress.

Ultimately the objectives of new technologies are: increased agriculture production, availability of drinking water to all the people of the earth, sufficient electric power generated to meet the needs of the industries and people, employment to all, protection from natural disasters as well as terrorism and violence, affordable healthcare and above all secured life to the people. Technologies must ensure sustained GDP growth rate with very low percentage of people below the poverty line and near 100 per cent literacy and strong defence, leading to a knowledge society.

In the Indian context it is essential to make the nation a developed country in line with the above objectives. Hence the Technology Vision 2020 was evolved by 500 scientists working for two years. This vision gives a road map in all technology disciplines for achieving higher economic growth for the country. In addition we are attempting to identify ten Futuristic Technologies which will have significance in the emerging world to have a position as one of the leading nations.

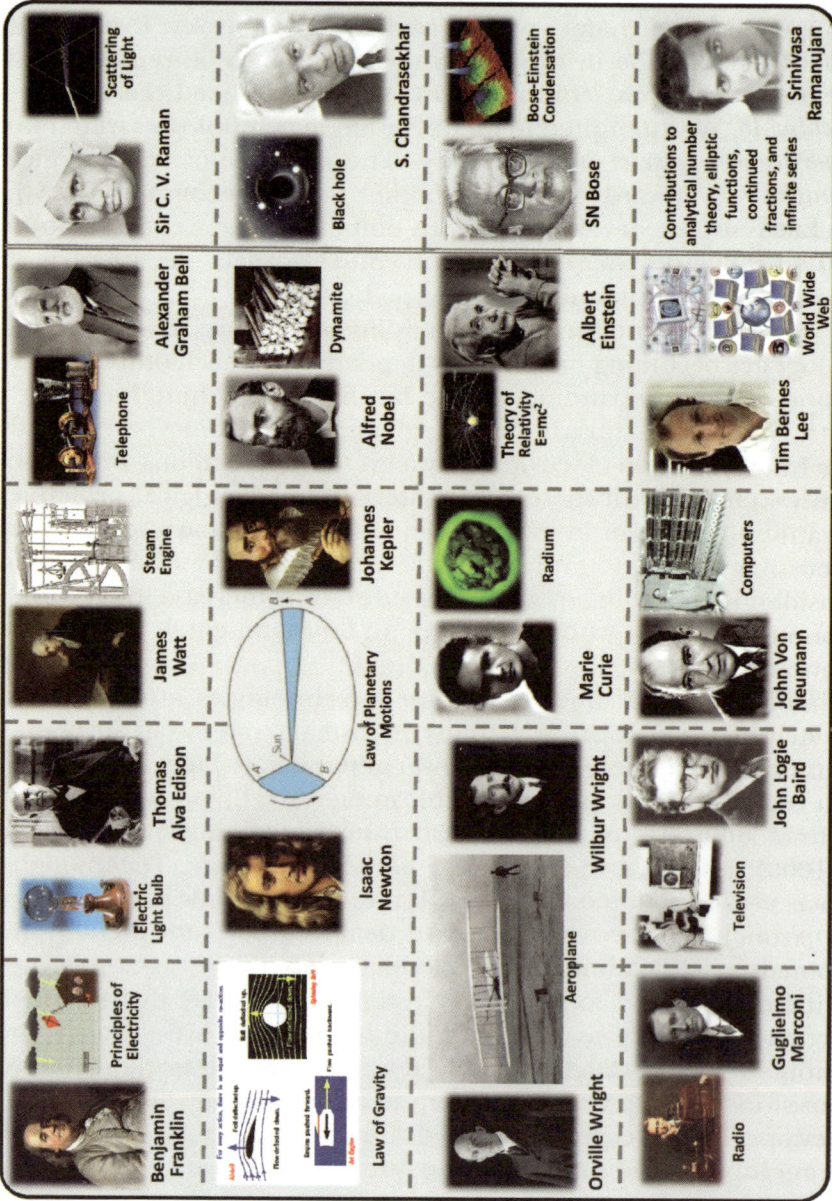

Fig. 3.2: Great Inventions and Inventors

Part–3 enumerates the advances made in (a) ICT, Bio and Nanotechnologies and their convergence (b) Robotics, Artificial Intelligence and Cognitive Sciences (c) Sensor Technologies (d) Materials (e) High Energetics (f) Fusion Technology and Green technology. These technologies provide essential growth for the nation in geometric progression.

In the subsequent parts, futuristic technologies which are essential for reaching the expected world standard are discussed. These are Hypersonics, Geo-Spatial Technologies, Low cost access to Space for generating energy through satellite and Mining in planets for high energy materials.

Each of these technologies is presented with a brief description, status, trends and reference to the progress in India.

3.2 ICT, BIO AND NANOTECHNOLOGIES AND THEIR CONVERGENCE

3.2.1 Information and Communication Technology (ICT)

The Information Age, also commonly known as the Digital Age, is the ability of individuals to transfer information freely, and to have instant access to information that would have been difficult or impossible to obtain previously. This digital revolution has considerable ramifications of a shift from traditional industry to an economy based on the manipulation of information, i.e., knowledge economy. Revolution in the computers and communication technology has resulted in the rapid conversion to knowledge society.

According to Moore's Law, computer chips (processors, memory, etc.) double their complexity every two years. This means that every fifteen years, on an average, a large number of technological capacities (memory, input, output, processing) grow by 1000X (Ten doublings: 2, 4, 8.... 1024). The capabilities such as processors, memory capacity sensors etc. of many electronic devices are strongly linked to Moore's law. All of these are improving at exponential rates. This is dramatically enhancing the impact of digital electronics in nearly every segment of the world economy. Recently, Intel unveiled a new microprocessor based on 22 nanometre process technology, the first high-volume chip to use 3-D transistors, and packs almost 3 billion of them onto a single circuit. They operate at much lower voltage and lower leakage, providing an unprecedented combination of improved performance and energy efficiency.

Similarly, the data storage has shown a revolutionary start with the 5¼ inch floppy drive and later with a 3½ inch one. Although smaller in capacity, they were extremely lightweight and portable and easy for storage and access. Data storage continued to make exponential progress into the 1990s and beyond. Floppy disk was replaced by Compact Disk which has the storage capacity of 700 MB, which in turn was replaced by DVDs having the storage capacity of 4 GB. Nowadays the storage media comes in the form of external hard disks with 1 TB capacity avoiding the risk of loss of data in case of the optical media like CDs / DVDs. Trends have consistently shown exponential growth in this area.

The Internet has enabled people to connect with the rest of the world and share information in the form of instant messaging, e-mails, social networking and online shopping or e-banking or e-governance. The revolution in communication technology has made the Internet a very popular tool in the exchange of data. According to the Internetworldstats.com, more than 2.4 billion, i.e; people nearly one third of the earth's population use the services of the Internet. National Science Foundation predicts that the number of Internet users will reach almost 5 billion by 2020. Today, the Internet has become the ultimate platform for accelerating the flow of information and is the fastest growing form of media.

Information Dominance has undergone three major changes. During the infancy stages of Information Technology evolution,

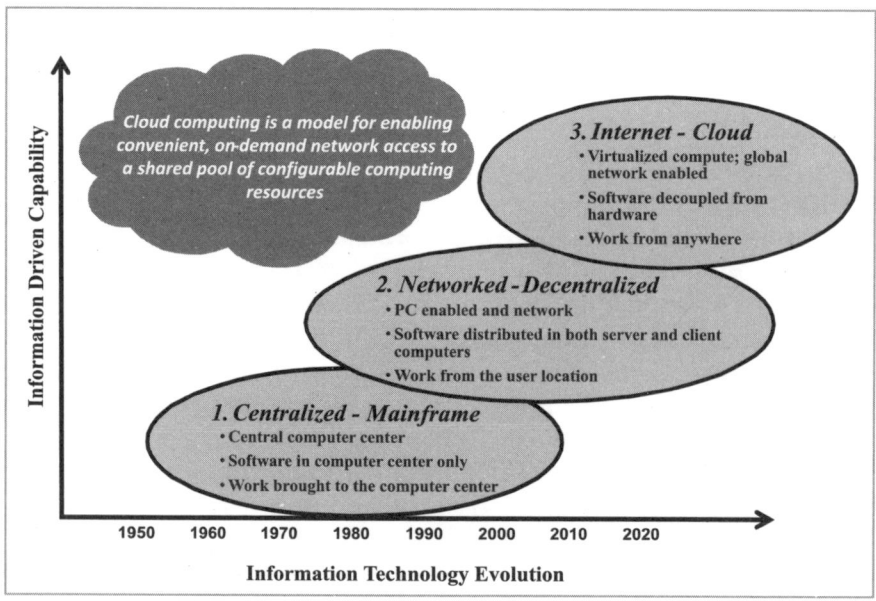

Fig. 3.3: Information Dominance

everything was centralized with central computer or mainframe with software available on the central computer. All tasks were carried out on the central computer. Information dominance shifted to the second level of networked systems which decentralized the tasks enabling individual PCs with network. Software was distributed on the server to the client and the work was carried out at the User node itself. The new term that is now occupying the stage is Cloud Computing. Cloud computing is the next generation Internet computing. With the use of global networking of systems, virtual computing has become popularized. Software is decoupled from hardware and users can work from anywhere irrespective of the location using the software available on the global network. According to National Institute of Standards and Technology, Cloud computing is a model for enabling convenient, on-demand network access to a shared pool of configurable computing resources like networks, servers, storage, applications, and services (Figure 3.3).

Cloud computing offers three types of services. They are Software as a Service (SaaS), Platform as a Service (PaaS) and Infrastructure as a Service (IaaS). SaaS is a Software deployment model whereby a provider licenses an application to customers for use as a service on demand. Examples of SaaS are Internet Services, Blogging / Surveys/ Twitter, Social Networking, Information / Knowledge Sharing, ERP etc. Optimized IT and developer tools are offered through Platform as a Service (PaaS) for Database and Testing Environments. Examples of PaaS are Application Development, Database Management, Directory Services etc. Infrastructure as a Service (IaaS) is on-demand like highly scalable Computing, Storage and Hosting Services with Mainframes, Servers, Storage, IT Facilities/Hosting Services etc. The main advantage of these services of Cloud Computing is that the organizations need not invest on computer resources, infrastructure and administration instead they can pay for the resources they have consumed through cloud computing.

Bringing about a fast evolution of technology in daily life, as well as in educational setup, the Information Age has allowed rapid global communication and networking to shape modern society.

In communication technologies the networks are based on physical media using glass fibres with optical wave guide, copper media with co-axial cables and wireless transmission with wimax or wi-fi.

The Communication network enables transmitting many applications and services simultaneously and on the same physical network. Modern communication technology has a life cycle of 3-10 years.

4G Technology

The third generation (3G) communication systems aims at providing faster communications services, entailing voice, fax and Internet data transfer capabilities anytime, anywhere throughout the globe, with seamless roaming between standards. The 3G technology supports around 144 Kbps, with high speed movement, 384 Kbps locally, and up to 2Mbps for fixed locations.

Need for 4G

In order to meet the needs of future high-performance applications like multi-media, full-motion video, wireless teleconferencing, packet switching network and increase in speed to 100Mbps for a moving user and 1Gbps for a stationary user, 4G technology is essential.

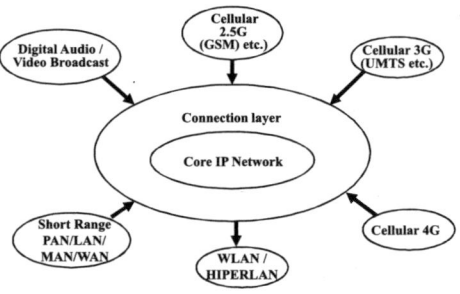

Seamless Connection of Networks in 4G

4G is about convergence of wired and wireless networks, wireless technologies including GSM, wireless LAN, and Bluetooth as well as computers, consumer electronics, communication technology and several others. 4G is a Mobile multimedia, anytime anywhere, Global mobility support, integrated wireless solution, and customized personal service network system. The seamless connection of networks in 4G.

The 4G technology is also referred to as "MAGIC" which stands for Mobile multimedia, Anywhere, Global mobility solutions over Integrated wireless and Customized services.

5G Technology

Beyond 4G, one can think of a real wireless world with no more limitations and access as well as zone issues, wearable devices with Artificial Intelligence (AI) capabilities, internet protocol version 6 (IPv6), where a visiting care-of mobile IP address is assigned according to location and connected network, one unified global standard and passive networks providing ubiquitous computing, allowing different radio technologies, including cognitive radios to share the same spectrum efficiently in the 5G technology.

5G technology could offer high resolution for cellphone users and bidirectional large bandwidth shaping, subscriber supervision tools for fast action, provision for automatic error avoiding, advanced billing interfaces, large broadcasting of data in gigabit supporting more than 65000 connections, transporter class gateway, remote diagnostics, virtual private network and so on. Revolution of 5G technology is about to begin. There are lot of improvements systematically from 1G, 2G, 3G and 4G to 5G in the world of telecommunications. A comparison of 1G to 5G technologies is given in the table. Effectively 5G will offer high speed, high capacity and low cost per bit; interactive multimedia, voice, streaming video, internet and other broadband services; more effective and more attractive with accurate traffic status; global access, service portability and scalable mobile services. Migration to 5G networks will ensure convergence of technologies, applications and services in a flexible platform in wireless with better return on investment and operating efficiency. This will change the lives of our people.

Technology/ Features	1G	2G/2.5G	3G	4G	5G
Start/ Deployment	1970/ 1984	1980/ 1999	1990/ 2002	2000/ 2010	2010/ 2015
Data Bandwidth	2 kbps	14.4-64 kbps	2 Mbps	200 Mbps to 1 Gbps for low mobility	1 Gbps and higher
Standards	AMPS	2G: TDMA. CDMA. GSM 2.5G: GPRS, EDGE. lxRTT	WCDMA. CDMA-2000	Single unified standard	Single unified standard
Technology	Analog cellular technology	Digital cellular technology	Broad bandwidth CDMA. IP technology	Unified IP and seamless combination of broadband, LAN/WAN/ PAN and WLAN	Unified IP and seamless combination of broadband. LAN/WAN/PAN /WLAN and wwww
Service	Mobile telephony (voice)	2G: Digital voice. short messaging 2.5G: Higher capacity packelized data	Integrated high quality audio, video and data	Dynamic information access, wearable devices	Dynamic information access, wearable devices with AI capabilities
Multiplexing	FDMA	TDMA, CDMA	CDMA	CDMA	CDMA
Switching	Circuit	2G: Circuit 2.5G: Circuit for access network & air interface; Packet for core network and data	Packet except circuit for air interface	All packet	All packet
Core Network	PSTN	PSTN	Packet network	Internet	Internet
Handoff	Horizontal	Horizontal	Horizontal	Horizontal and Vertical	Horizontal and Vertical

Information and Cyber Security

In a networked scenario, there is no assurance of safety—if it is networked, then it is available to all others on the network. Economies of today are more and more dependent on information and communications technology (ICT) thereby becoming more vulnerable to network attacks. The most severe cyber security risks are those that threaten the functioning of critical information infrastructures, such as those dedicated to financial services, control systems for power, gas, drinking water, and other utilities; airport and air traffic control systems; logistics systems; military operations and government services.

Although the prevailing basis for cyber security is to ensure a favorable climate for ICT, national and international security concerns are becoming equally important grounds. Cyber security is thus a collective concern that is comprehensive in scope—the Internet has no national boundaries. The capacity for cyber risk management and security lies largely in the hands of private entities that manage and operate most ICT infrastructure. Such security cannot be adequately assured by market forces rather it requires a combination of solutions involving a range of stakeholders working both in their own domains and in concert.

With the spread of high-speed Internet infrastructure new risk factors and challenges to data and communications networks are rapidly emerging. Computer worms and viruses, organised criminal activity, weak links in the global information infrastructure, hacker-activists, potential military operations, rapid evolution of information and communication technologies are among the few risks.

3.2.2 Bio Technology (BT)

Biotechnology involves techniques that use living organisms or substances from those organisms to make or modify a product, improve plants or animals, develop micro-organisms for specific uses or to produce new products and new forms of organisms. The impact of the biotechnology related developments in agriculture, health care, environment and industry has already been visible and the efforts are now culminating into products and processes. In India, more than a decade of concerted effort in research and development in identified areas of modern biology and biotechnology has given rich dividends. Patenting of innovations, technology transfer to industries and close interaction with them has given a new direction to biotechnology research. Initiatives have been taken to promote transgenic research in plants with emphasis on pest and disease resistance, nutritional quality, silk-worm genome analysis, molecular biology of human genetic disorders, brain research, plant genome research, development, validation and commercialisation of diagnostic kits and vaccines for communicable diseases, food biotechnology and biodiversity conservation etc. The applications of biotechnology are majorly in four areas – agriculture, health care, industrial uses of crops like biodegradable plastics, bio-fuels etc. and environmental uses like bioremediation.

BT in Agriculture

Agriculture is the backbone of most developing countries, with more than 60 per cent of the population reliant on it for their livelihood. Nanotechnology can also improve our understanding of the biology of different crops and thus potentially enhance yields or nutritional values. By transferring one or two genes to develop a high yield crop imparting new character, it could result in the increase of the total agricultural yield. The country is facing problems of frequent drought conditions, low temperature spells and a number of salt affected areas. There is need to search for genes to overcome these problems by developing suitable crop varieties. Genetic engineering techniques would enable the new crop varieties with better withstanding capabilities against vulnerable stresses like salt, drought, cold and heat. The pest resistant genes for various biotic stresses can be a big boon to the farmers and substantially boost agriculture production. Population in a country like India suffers from deficiencies of Vitamin A, Iodine, Calcium and Iron. This is particularly true for women in general and rural populace in particular leading to unexpected levels of infant mortality rate and maternal mortality rate. Lack of supplementation of diet with these nutrients is a problem. Modifying the proteins in food would result in improved nutritional qualities. If crops or crop varieties rich in Vitamin A, Iodine, Calcium and Iron can be developed by incorporating appropriate genes, it will enable improvement of our human development index of the nation. Modern biotechnology methods would slow down the spoilages and result in enhancing the shelf life of the vegetables for reasonable time. Value addition is vital for the agriculture produce to ensure adequate return to the growers. Genes can play a very big role in value addition to agriculture crops particularly vegetables, fruits, flowers and other perishables, so that, they reach the market and kitchen which is substantially far off from the farm. This value addition could be in the form of extension of shelf life by exploiting appropriate genes. This when combined with agricultural food processing can save the nation 25-35 per cent of the crop losses in Horticulture. Moreover, the Indian food chain had health as an important component in herbal medicines like turmeric, neem, pepper, garlic etc., but due to the scientific process of evolving allopathic medicine facing extreme competition in India, our medicinal plants have resurfaced from the Western source through systematic scientific processing, packaging and marketing. The study of these plants and genomic signature by Biotechnologists would enable us to patent these plants for medicinal use with adequate protection of intellectual property rights.

Agricultural Biotechnology

It refers to the use of scientific techniques to improve plants, animals and microorganisms. Scientists have developed solutions to enhance agricultural productivity. Starting from the ability to identify genes that may confer advantages on certain crops, and the ability to work with such characteristics very precisely, biotechnology enhances the breeders' ability to make improvements in crops and livestock. Agricultural biotechnology deals with genetic engineering, molecular markers, vaccines and tissue cultures (Figure 3.4).

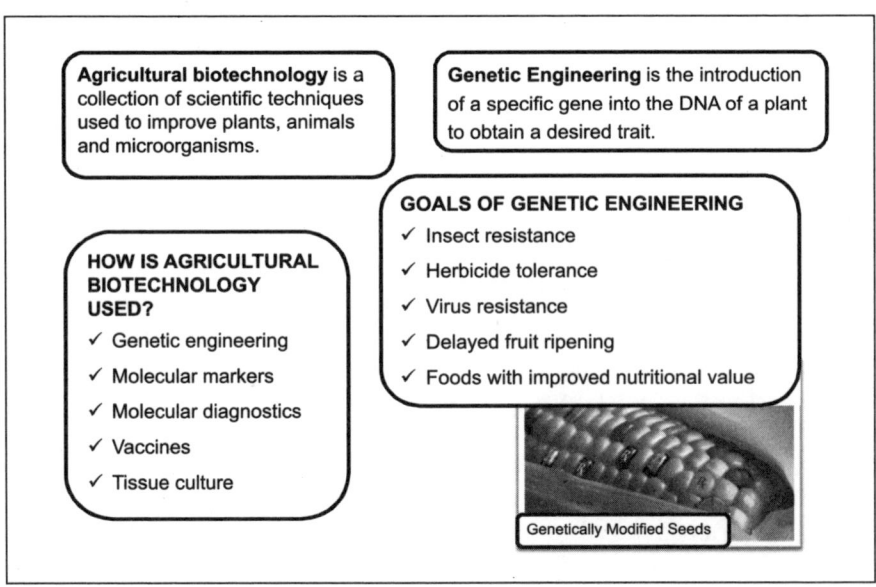

Agricultural biotechnology is a collection of scientific techniques used to improve plants, animals and microorganisms.

Genetic Engineering is the introduction of a specific gene into the DNA of a plant to obtain a desired trait.

HOW IS AGRICULTURAL BIOTECHNOLOGY USED?
✓ Genetic engineering
✓ Molecular markers
✓ Molecular diagnostics
✓ Vaccines
✓ Tissue culture

GOALS OF GENETIC ENGINEERING
✓ Insect resistance
✓ Herbicide tolerance
✓ Virus resistance
✓ Delayed fruit ripening
✓ Foods with improved nutritional value

Genetically Modified Seeds

Fig. 3.4: Agricultural Biotechnology

1) *Genetic Engineering:* Scientists have learned how to move genes from one organism to another. The process is called genetic modification (GM), genetic engineering (GE) or genetic improvement (GI). This process allows the transfer of useful characteristics (such as resistance to a disease) into a plant, animal or microorganism by inserting genes (DNA) from another organism. Virtually all crops improved with transferred DNA to date have been developed to aid farmers to increase productivity by reducing crop damage from weeds, diseases or insects.

Goals of Genetic Engineering:

a) *Insect resistance:* In the last few years, several crops have been genetically engineered to produce their own Bt (Bacillus thuringiensis is a soil bacterium that contains a protein that is toxic to a narrow range of insects, but harmless to animals or humans) proteins, making them resistant to specific groups of insects. Applications of Bt bacteria have been used to control insect pests for many years before the advent of the current Bt crops that have been made using biotechnology.

Varieties of BT insect-resistant corn and cotton are now under commercial production. Other crops being investigated include cowpeas, sunflower, soybeans, tomatoes, tobacco, walnut, sugar cane, and rice.

b) *Herbicide tolerance:* Weeds growing in the same field with crop plants can significantly reduce crop yields because the weeds compete for soil nutrients, water, and sunlight. Many farmers now control weeds by spraying herbicides directly onto the crop plants. Because these herbicides generally kill only a narrow spectrum of plants, farmers apply mixtures of multiple herbicides to control weeds after the crop has begun to grow.

c) *Virus resistance:* Many plants are vulnerable to diseases caused by viruses which are often spread by insects (such as aphids) from plant to plant across a field. The spread of viral diseases can be very difficult to control and crop damage can be severe. The most effective methods against viral diseases are cultural controls (such as removing diseased plants) or plant varieties bred to be resistant to the virus but such strategies may not always be available. Where options were limited earlier, scientists have now discovered new genetic engineering methods that provide resistance to viral disease.

d) *Delayed fruit ripening:* Delaying the ripening of fruit is of interest to producers because it allows more time for shipment of fruit from the farmer's fields to the grocer's shelf, and increases the shelf life of the fruit for consumers. Fruit that is genetically engineered to delay ripening when left to mature on the plant longer, it will have longer shelf-life in shipping, and may last longer for consumers.

e) *Foods with improved nutritional value:* Genetic modification can be used to produce crops that contain higher amounts of vitamins in order to improve their nutritional quality. For example, genetically changed "golden rice" contains three transplanted genes that allow plants to produce beta-carotene (a compound that is converted to vitamin A within the human body). Deficiency of Vitamin A is the world's leading cause of blindness—it affects as many as 250 million children.

Genetic engineering has also been used to alter the content of many oil crops, either to increase the amount of oil or to modify the types of oils they produce. Biotechnology could also be used to upgrade some plant proteins of low biological value because they lack one or more of the 'essential' amino acids. Examples include maize with improved protein balance and sweet potatoes with increased total protein content. Reducing toxicity of certain foods is also a goal of biotechnology. For example, reduction of the toxic cyanogens in cassava has been shown to be possible and could be produced in the future.

2) *Molecular markers:* By examining the DNA of an organism, molecular markers can be used to select plants or animals that possess a desirable gene, even in the absence of a visible attribute. Thus, breeding is more precise and efficient. For example, the International Institute of Tropical Agriculture has used molecular markers to obtain cowpea resistant to bruchid (a beetle), cassava resistant to Cassava Mosaic Disease and disease resistant white yam, among others. Another use of molecular markers is to identify undesirable genes that can be eliminated in future generations.

3) *Molecular diagnostics:* Molecular diagnostics are methods to detect genes or gene products that are very precise and specific. Molecular diagnostics are used in agriculture to diagnose crop/livestock diseases more accurately.

4) *Vaccines:* Biotechnology-derived vaccines are used in livestock and humans. They may be cheaper, better and/ or safer than traditional vaccines. They are also stable at room temperature, and do not need refrigerated storage; this is an important advantage for smallholders in tropical countries. Some are new vaccines which offer protection for the first time against some infectious illnesses. For example, in Philippines, vaccine is used as a preventive measure to protect cattle and water buffalo against hemorrhagic septicemia, a leading cause of death in both species.

5) *Tissue culture:* The process of regeneration of plants in the laboratory from disease-free plant parts is known as tissue culture. This technique allows for the reproduction of disease-free planting material for crops. Examples of crops produced using tissue culture include citrus, pineapples, avocados, mangoes, bananas, coffee and papaya.

BT in Healthcare

In healthcare, modern biotechnology has a vital role in applications like drug delivery, pharmacogenomics, gene therapy and genetic testing. Pharmacogeneomics is the study of pharmaceuticals, genetics and the body's response to drugs. Bioprocessing technology is one of the oldest form of processing technologies that uses microorganisms such as yeast, bacteria or the molecular components of their manufacturing machinery to produce desired products. Genetically altered microorganisms like E.coli or yeast are used in the production of synthetic insulin or antibiotics. Bioprocessing technology also encompasses tissue engineering and manufacturing as well as biopharmaceutical formulation and delivery. Today, recombinant DNA technology is used, coupled with microbial fermentation to manufacture a wide range of bio based products including human insulin, the hepatitis B vaccine, the calf enzyme used in cheese-making, biodegradable plastics, and laundry detergent enzymes.

Bio-Fuel

Bio-fuels are made from algae, sugar, starch, corn and oil. Biodiesel is the most common biofuel produced from oils or fats using trans-esterification and is similar in composition to fossil diesel. Jatropha seed is a common source of biofuel. Out of nearly 60 million hectares of wasteland, 30 million hectares are available for energy plantations like "Jatropha". Once grown, the crop has a life of fifty years. Under the optimum conditions, a yield of approx. 5 tonnes per hectare of oilseeds is possible. Output of Biodiesel from Jatropha seeds is approximately 2 tonnes per hectare that could be used in automobiles, and other agro-industrially useful byproducts. The cost of bio-diesel produced from Jatropha works out to Rs.20 per litre at present. Government has permitted the use of mixing of maximum 10 per cent bio-fuel with diesel for Cars. 100 per cent usage of bio-fuel has shown satisfactory results with respect to heavy vehicles and tractors. Biodiesel is carbon neutral and many valuable by-products flow from this agro-industry. Research focus has to be on developing high yielding varieties with a capacity to produce 4 to 6 kg seed yield per plant or 4 to 6 tonnes per hectare per annum.

Jatropha grows in a wide range of climatic conditions, and also exhibits a wide range of features in different climates. The variation is believed to be mostly climate-driven coupled with available soil water moisture. Hence research initiatives should be aimed at the production of flowering and fruiting with minimal moisture content in the soil. Large scale genomic studies are now being routinely performed in many laboratories all around the world. Extensive research has been conducted on plant genomes. Genomic sequencing of Jatropha would pave the way for elucidation of the relationship between genome evolution and phenotypic diversity. Moreover, it will enable us to search for genes and develop molecular markers associated with agricultural traits, thereby establishing a molecular breeding system contributing economically to the improvement of the Jatropha species.

Bioethanol is a form of renewable energy that can be produced from agricultural feed-stocks. It can be processed from common crops such as sugarcane, potato, manioc and corn. Production of ethanol from corn uses only a small part of the corn—the corn kernels taken from the corn plant and the starch, which represents about 50 per cent of the dry kernel mass, is transformed into ethanol by microbial (yeast) fermentation of sugars, distillation, dehydration or denaturing. Ethanol fuel is widely used in Brazil and in the United States, and together both countries were responsible for 87.1 per cent of the world's ethanol fuel production in 2011.

Another alternate to the fossil fuel is Algae fuel. Algae (macro and microalgae) usually have a higher photosynthetic efficiency than other biomass producing plants. Algae can produce up to three hundred times more oil per acre than conventional crops such as corn, rapeseed, palms, soybeans, jatropha. As algae have a harvesting cycle of 1–10 days, it permits several harvests in a very short time frame, a strategy differing from yearly crops. Algae can also be grown on land that is not suitable for other established crops, for instance, arid land, land with excessively saline soil, and drought-stricken land. Algae cultivation can be through Photobioreactors, closed loop system or Open pond. It requires sunlight, water and carbon dioxide to grow and can be produced using ocean and wastewater. It does not affect the fresh water resources. Various methods such as mechanical extraction using hydraulic or screw, enzymatic extraction, chemical extraction through different organic solvents, Ultrasonic extraction, and CO_2 supercritical extraction are available for the extraction of algal oil. Conversion of algal oil into biodiesel happens by base catalyzed transesterification with alcohol since micro algae has got very high oil yield to an extent of 19000 to 57000 litres per acre. It is estimated that 2-3 per cent of the total Indian cropping land is sufficient to produce enough biodiesel to replace all petrodiesel currently being used in the country.

Environment Protection by Bio-degradation

Biotechnology has tremendous applications in the protection of environment. Marine pollution due to petroleum products such as high speed diesel is a big threat to the marine flora and fauna. Natural dispersion of oil is a slow process which takes months. Biodegradation of hydrocarbons through bioemulsifier, nutrient and bacteria plays a vital role in oil dispersion. Naval Materials Research Laboratory of DRDO, has developed a bioremediation process. The bioremediation of pollutant oil is carried out by spraying bio-emulsifier, oil degrading marine bacteria and nutrient mixture at polluted site in intervals of three days for complete removal of oil (Figure 3.5).

The oil hydrocarbons are degraded by bacterial alkane monoxygenases and associated enzymes releasing CO_2 and water and no toxic substance. The other bio-molecules generated get incorporated as biomass and 'Bio-emulsifier' as byproducts. The bio-emulsifier developed is a heat stable compound having a molecular weight of ~5000 as determined in gel permeation chromatography (GPC). Bio-emulsifier is a surface active biomaterial which facilitates emulsification of oil-in-water, enhancing bioavailability of oil for biodegradation in an eco-friendly manner.

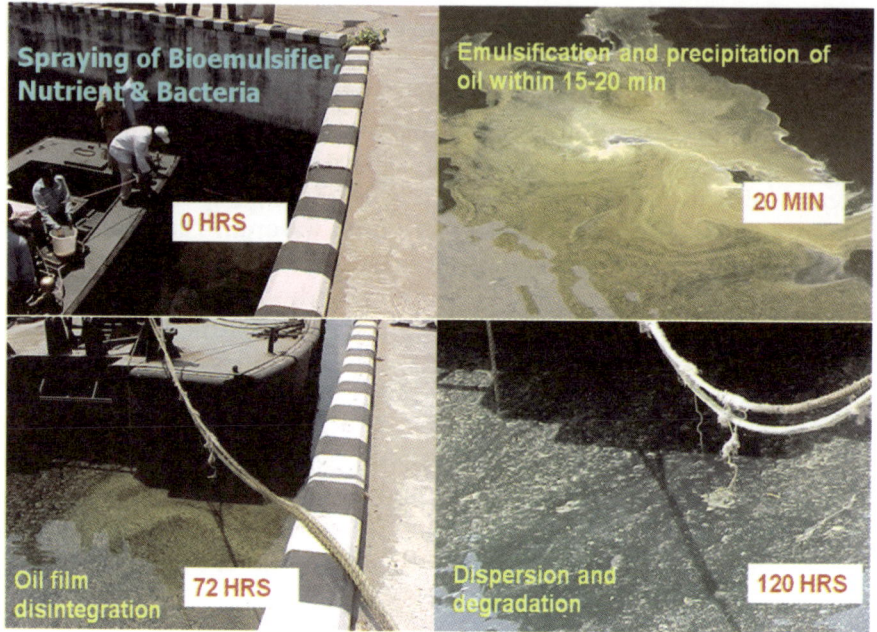

Fig. 3.5: Bioremediation of Floating Oil

Bio-informatics

Application of Information technology in the field of Biotechnology results in bioinformatics. Bioinformatics has an extensive use in the field of biological research through the management and analysis of data using advanced computational techniques. Bioinformatics is associated with the generation of databases in genomics and proteomics which in turn are associated with the information of genes and proteins. Bioinformatics plays a key role in creation and maintenance of a database to store complex data from biological processes. This helps in the sequencing of DNA. Genome annotation is the process of marking genes and other biological features in DNA sequencing. For this purpose, there are many softwares designed for sequencing and analysis of genome of living organisms. The advent of Information technology revolution will definitely pave way for new software tools and techniques in the application of Biotechnology. Department of Biotechnology has established many Bioinformatics centres that disseminate bioinformatics information to the scientists in R&D laboratories and Academia. Information technology which helps in the form of networking of these centres through satellite communication system encouraging sharing of information along the country could be accessed for the benefit of the countrymen.

Artificial Neural Network Out of DNA

Recently, the Researchers of California Institute of Technology created the first Artificial Neural Network out of DNA by creating a circuit of interacting molecules that can recall memories based on incomplete patterns, in the same way as can a brain (Figure 3.6). The neural network was built of four neurons made up of 112 distinct DNA strands (by contrast, the human brain has some 100 billion neurons). This DNA-based neural network demonstrates the ability to take an incomplete pattern and figure out what it might represent—one of the brain's unique features. According to the researchers, biochemical systems with artificial intelligence (or at least some basic, decision-making capabilities) could have powerful applications in medicine, chemistry, and biological research in answering fundamental biological questions or in diagnosing a disease. Duplication of Human Brain is not far with the nano revolution.

Simulation of Human Brain

The Human Brain Project of European Union intends to create a computer simulation of the 89 billion neurons inside our skull and the 100 trillion connections that wire those cells together. This will become a virtual copy of the human brain enabling basic research on

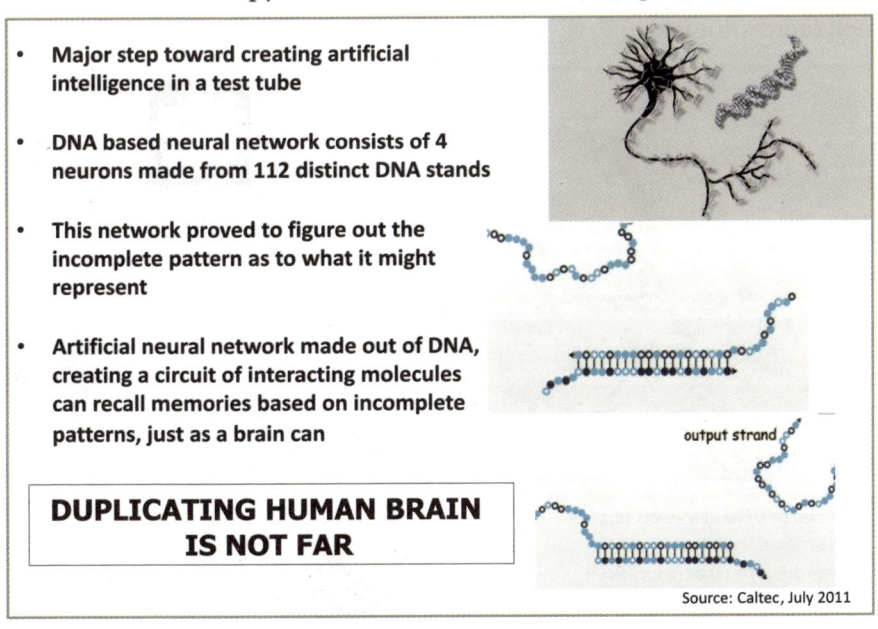

- Major step toward creating artificial intelligence in a test tube

- DNA based neural network consists of 4 neurons made from 112 distinct DNA stands

- This network proved to figure out the incomplete pattern as to what it might represent

- Artificial neural network made out of DNA, creating a circuit of interacting molecules can recall memories based on incomplete patterns, just as a brain can

output strand

DUPLICATING HUMAN BRAIN IS NOT FAR

Source: Caltec, July 2011

Fig. 3.6: Development of First Artificial Neural Network out of DNA

brain cells and circuits. This may also lead to computer based drug trials. This project of 1 bn Euro from the European Union is expected to model each level of brain function from chemical to electrical signaling up to the cognitive traits that underlie intelligent behaviour. The ability to simulate the brain in enough detail will demand computing memory in exabyte (10^{18}) and computing speed in exaflops.

Stem Cell

Worldwide stem cell research has gained momentum. The result of stem cell research is outstanding in the treatment of dreadful diseases like cancer, nerve disorders, cardiac disease, etc. Billions of dollars are being invested by USA, Europe, Japan, South Korea, Canada, Australia and other countries on stem cell research. India is also keenly pursuing stem cell research. Institutions in collaboration with academia and industry are pioneering research activities. Presently, in India, over thirty institutions and hospitals are engaged in this research. A large number of research facilities have been set up and are being networked with leading hospitals and labs such as CCMB, AIIMS, LVPEI, IISc, Reliance Life Sciences, etc., with the support of DBT, government labs and industries. The market is expected to grow to US$500 billion by 2020 for regenerative medicines and stem cell technologies.

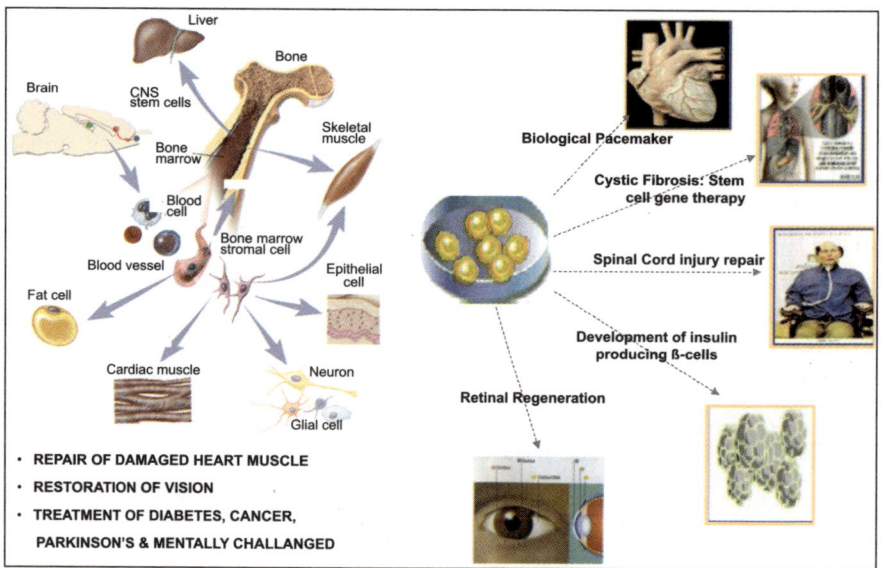

Fig. 3.7: Stem Cells Applications

Stem Cell Applications

The ultimate aim of bio scientists and technologists is to rebuild or replace injured or diseased tissues/organs that cannot heal naturally, through stem cell research. This would be accomplished by transplanting stem cells into the damaged area and directing them to grow a new healthy tissue. Applications of stem cell therapy include gene therapy, repair of damaged heart muscle, restoration of vision, treatment of diabetes, cancer, Parkinson's disease, nerve disorder, spinal cord injury repair, etc. (Figure 3.7). Aarohi a four-month-old child who was suffering from Alcapa (anomalous origin of left coronary artery from pulmonary artery) underwent stem cell therapy at Frontier Lifeline Hospitals, Chennai. This therapy was unique and the first of its kind in the country.

Limbal Tissue Transplantation for Cornea at LVPEI

Limbal tissue transplantation for cornea has been successfully done at LV Prasad Eye Institute. Limbal stem cells are produced in the cornea and they regenerate the ocular surface of the cornea. Limbus damage due to burn or injury leads to stem cell deficiency which causes corneal scarring and blindness. The damaged limbal stem is replenished through the stem cells extracted and transferred from the limbus of patient / relative followed by corneal transplantation. So far, 125 patients have been successfully treated in LVPEI, Hyderabad.

Heart Repair through Stem Cell

The term 'heart failure' means the heart is not working as efficiently as it should. One of the reasons of heart 'failure' is the damage of the heart muscle. It is not a disease but rather a 'syndrome' which may arise from a number of causes. Stem cells are multi-potential, self-renewing cells and their growth and differentiation into specific cells in order to repair the heart muscle holds the key. Healthy cells are grafted and transplanted into the damaged portion of the heart to repair it. The sources of cells could be skeletal myoblasts, embryonic stem cells, smooth muscle cell, bone marrow stem cells or embryonic cardiomyocytes. This was successfully tried out at All India Institute of Medical Sciences, New Delhi.

Stem Cell Therapies for Neurological Disorders

Neurological disorders involve the loss of particular cell types in the

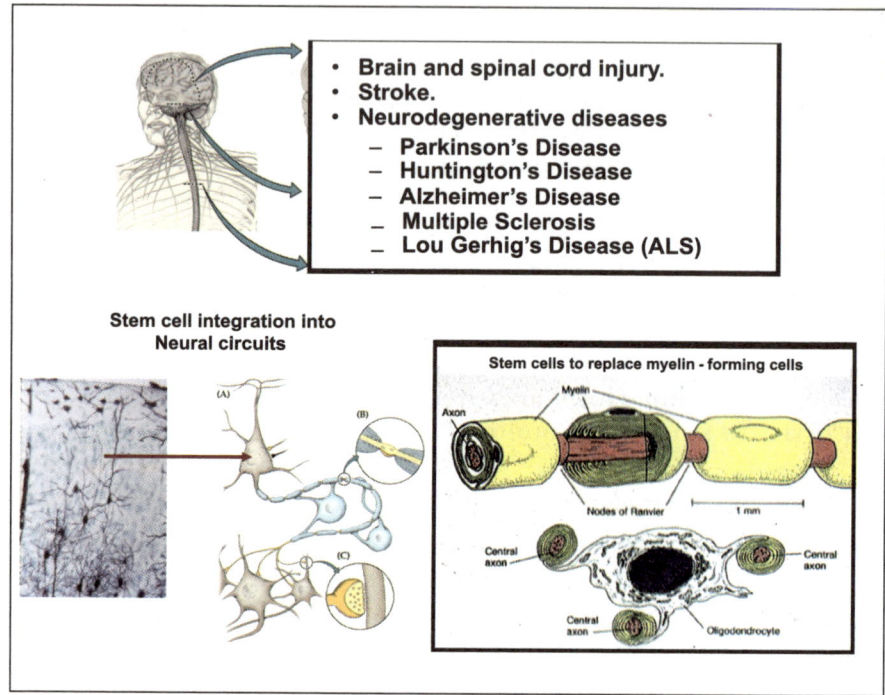

Fig. 3.8: Stem Cell Therapies for Neurological Disorders

nervous system. Embryonic stem cells are used to restore the cells that are lost as a result of injury or neurodegenerative diseases. The stem cells are converted to nerve cells by differentiation into neurons. Stem cells grow and integrate into neural circuits to overcome the neurological disorder (Figure 3.8).

Isolation and Perpetuation of Brain Tumour Stem Cell in Culture

Adult precursors are dissociated and plated in a liquid growth medium that contains the stem-cell mitogens. Because of the lack of serum and low plating density, most cells die except those that divide in response to stem-cell mitogens (Figure 3.9). The growth-factor-responsive cells proliferate to form floating clusters of cells that are referred to as neurospheres. The neurosphere assay is a serum-free culture system that allows the isolation and propagation of CNS-derived stem cells. These can be further dissociated into a single-cell suspension and then re-plated in a fresh medium to produce secondary neurospheres. The process can be repeated, resulting in a geometric expansion in the number of cells that are generated at

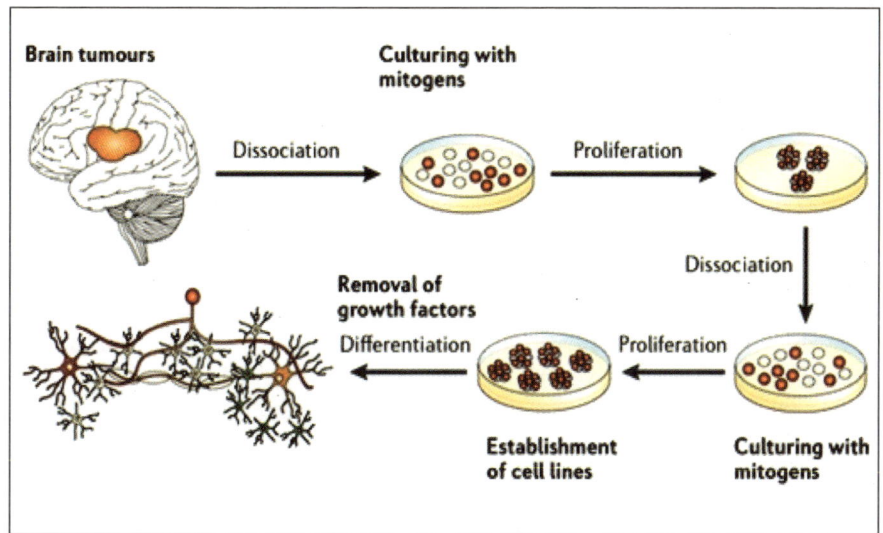

Fig. 3.9: Isolation and Perpetuation of Brain Tumour Stem Cells in Culture

each passage. Adult neural stem cells possess the fundamental stem-cell features of extensive self-renewal, generation of many progeny and the ability to give rise to the primary cell types of the tissue from which they were obtained.

Therapeutic Implications of Cancer Stem Cells

Conventional therapies may shrink tumours mainly by killing cells with limited proliferative potential. Cancer stem cells are less sensitive to these therapies. They remain active after therapy and regenerate the tumour. These therapies could be targeted against cancer stem cells so that the tumour is degenerated and unable to grow further. Thus, the cancer stem cell-directed therapies would not only degenerate the tumour and cure it but also lead to minimal injury to other healthy cells.

A recent breakthrough in cancer research at Stanford University School of Medicine by an Indian American researcher proves that in times of stress, certain human embryonic stem cells produce molecules that benefit themselves along with helping nearby cells to survive. A better understanding of human embryonic stem cells altruism could provide new insights into cancer therapies as well as the ability to develop safe and effective stem cell treatments for other diseases.

3.2.3 Nanoscience and Technology

Biological and Molecular Complexity

Even after many years, the colour of the peacock's feather does not fade away. This undying colour phenomenon of the peacock has come from God's own creation of nano particles spread in a peacock's feather as they diffract light which gives us rich colours. Everything in this universe from giant stars to our bodies works on a molecular scale. Our hearts and lungs are big objects but all the processes take place at a molecular level. Therefore, everything in our body and in the physical universe is already based on nanotechnology. Fabrication, manipulation and self-assembly of atoms, molecules into structures with a dimension of 1-100 nanometer to create materials with new and superior properties is nanoscience. At this scale, it opens up new dimensions of applications. The amalgamation of nanoscience and technology with nano manufacturing will bring new devices for applications in practically all fields e.g. material science, medical, electronics, chemical etc. Nanotechnology represents the most fundamental paradigm shift. Till now technology has been one way, from top to bottom, where we start with sizeable pieces of materials and work them down towards smaller and more precise dimensions. With dimensions becoming tinier, doing things has become harder. Nanotechnology is fortunately in the reverse direction from bottom up and the ability to build brand new substance properties through control and to manipulate substances on nanoscale. Nanotechnology is going to usher in a new technological revolution.

1 nm = 1 billionth of a meter = 10^{-9} m. Each nanometre is about 4 to 5 atoms wide and almost 80,000 times smaller than human hair. Nano does not merely mean smaller, it is basically putting atoms and molecules together in an ordered fashion to create remarkable mechanical, optical and chemical properties. Nanotechnology is the art and science of manipulating matter at the atomic or molecular scale. We believe that nanotechnology is a new technology that is knocking at our door. It has deeper reach compared to information technology and will touch the common man. It will be the central focus for many technologies to converge and open a variety of applications which are unimaginably large. If one sees the world nano publications scenario, we realize that India has a long way to go. Rapid advances both in terms of materials and devices are taking place globally. Many of these inventions in materials and devices that are being created today by nanotechnology were beyond human

imagination few years back. Further, this technology will have a large domestic market potential and hence would be very robust and immune to the changes that would take place beyond our borders.

Dimensions of Nanotechnology

Dimensions of nanoscience and technology are large and the development efforts towards this will have far reaching consequences in the areas of production and manufacturing leading to compact state-of-the-art systems. Benefits of this would be enjoyed by all sectors of society in the country. A well envisioned mixture of fundamental sciences and applied sciences, and a true application of nanosciences would lie at the interface of many competing and fully-grown disciplines such as biology, chemistry, physical sciences and engineering.

Molecular nanotechnology has enormous potential for future aerospace systems. Traditionally, microscopic devices have been made by being cut or formed from large objects. But as these products shrink below the micron level this process becomes increasingly difficult. Recently, chemists have begun to try the opposite approach, i.e., building nanoscale objects from molecular building blocks. Although these devices arc too small to manufacture by traditional material science approaches, they are also far too large to be synthesized by classical chemical synthesis. In order to reach these nanoscale devices upwards from a molecular level, a massively convergent synthesis is required.

Research has shown that most biomolecules are far too fragile for many aerospace environments. For example, it is unlikely that proteins or DNA can survive in rocket engines. Research has shown that a newly discovered class of molecules, Fullerenes, essentially carbon nanotubes built from graphite sheets curved into a wide variety of close shapes, may lead to tougher, high temperature materials that can survive in vacuum and other harsh environments. Fullerenes have certain advantages. Carbon nanotubes are a normal form of carbon with remarkable electrical and mechanical properties. Carbon nanotubes can be visualized as rolled-up graphite layers formed into cylinders. Carbon nanotubes and its composites will give rise to super strong, smart and intelligent structures in the field of material science. The future scenario for materials technology would include both intelligent and ultra-strong materials with much higher values of Young's Modulus.

Nanotechnology Applications

A well-structured composition of micro, macro and nanoporous materials will give rise to smart clothing. Embedded nanoparticles can be used to create stain repellent khakis resulting in savings on account of detergent cost. Smart clothing can be used by our defence personnel from Siachen to Rajasthan for their camouflaging. Molecular switches and circuits along with nanocell will pave the way for the next generation computers. Ultra dense computer memory coupled with excellent electrical performance will give the society low power, low cost, nanosize and yet faster assemblies. The last four decades have also affected the packaging concept. The whole concept of miniaturization is possible today because of the concept of packaging. Electronics packaging of the past has given way to the present Micro-systems packaging and the shift in the trend is now towards futuristic nanopackaging. Packaging industry can use nanoparticles as material to reduce UV exposure and subsequent spoilage.

CNT-based Solar Cells for Higher Efficiency

The era of wood and biomass is almost nearing its end. The age of oil and natural gas would soon be over within the next few decades. The world energy forum has predicted that fossil-based oil, coal and gas reserves will last only for another 5–10 decades. Nuclear, hydrogen fuel and solar rays are the three modes to get clean power.

Solar rays, when passed through presently available solar photovoltaic cells, have an efficiency of less than twenty per cent. The low efficiency of conventional photovoltaic cells has restricted the use of solar cells for large applications for power generation. Research has shown that the Gallium Arsenide (GaAs) based PV cell with a multi-junction device could give a maximum efficiency of 30 per cent. Therefore, the focus of present research is on the use of Carbon Nano Tube (CNT) based PV cell. Both single wall CNTs and multiwall CNTs have been used as electrodes and electron acceptor, which can split exciton into electrons and holes to produce electricity (Figure 3.10).

CNTs provide better electron ballistic transport property along its axis with high current density and capacity on the surface of the solar cell without much loss. Higher electrical conductivity and mechanical strength of CNT could improve the quantum efficiency. Research abroad has shown that the alignment of the CNT with the

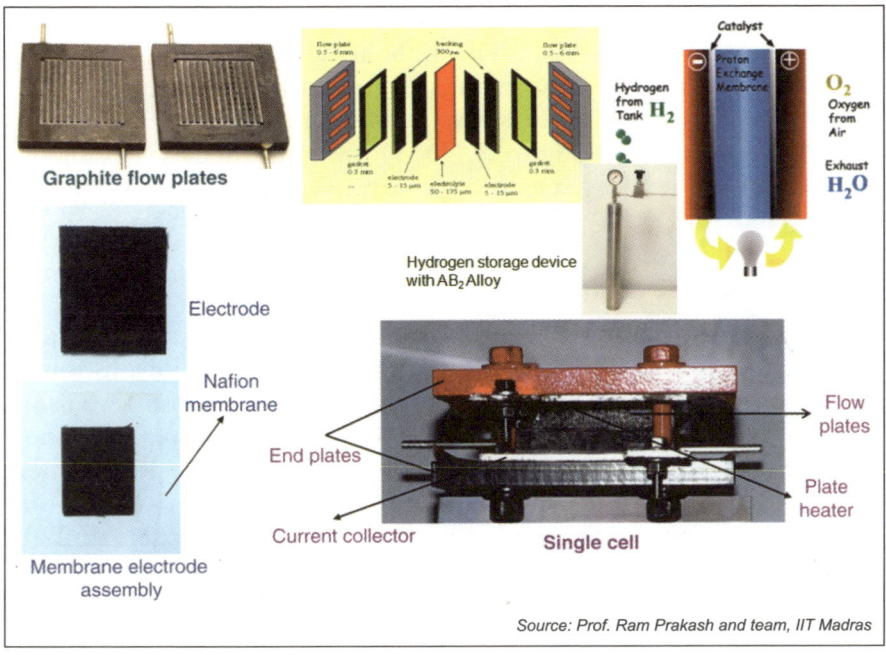

Fig. 3.10: MWNT Based Fuel Cell

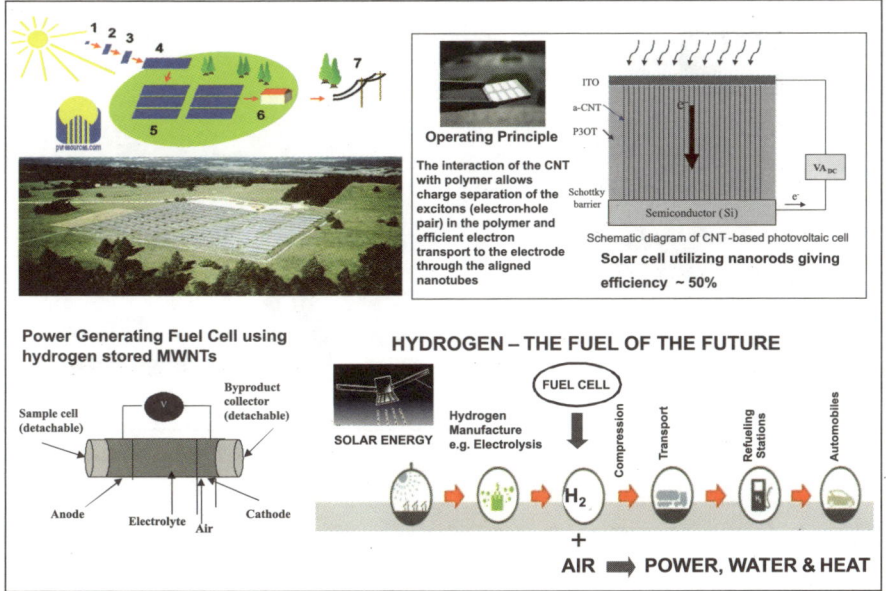

Fig. 3.11: Nanotechnology Application for Energy (Source: Prof. Varadan)

polymer composites substrate is the key to giving very high efficiency in photovoltaic conversion. The polymer composites increase the contact area for better charge transfer and energy conversion. In this process, researchers could achieve an efficiency of about 50 per cent at the laboratory scale (Figure 3.11). This will become a cost effective option for generating electric power, particularly in a tropical country like India.

Wearable Power for Soldier

The combination of nano-systems, nanoelectronics and nanomaterials will dominate the battle scene. This "Revolution in Nanotechnology" will soon become one of the most important concepts in strategic planning and may redefine the planning and deployment concept of future emerging battlefields. Researchers have developed technology that combines multiple materials into intricately structured fibres that can store energy or convert sunlight into power (Figure 3.12), for use in soldiers' uniforms. This helps in the reduction of weight for the soldier.

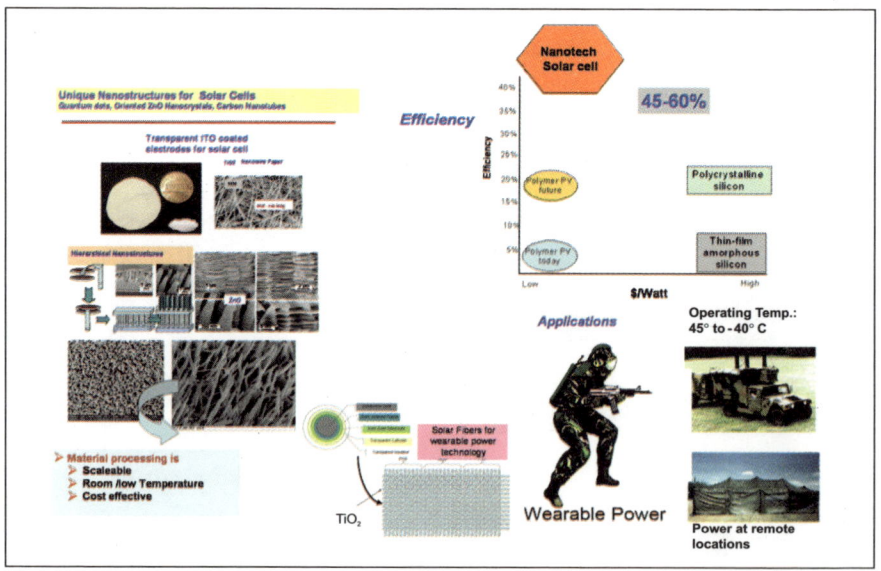

Fig. 3.12: Wearable Power Generation (Image Source: Prof. Varadan)

Nano Tube Filter for Water Purification

More than 70 per cent of the earth's surface is covered with water but only 1 per cent is available as fresh water for drinking purposes.

Today, out of a global population of 6 billion, only 3 billion have access to limited or perhaps satisfactory supply of water. It is estimated that 33 per cent of the world population has no access to sanitation and 17 per cent has no access to safe water. By 2025, the world population is going to rise to 8 billion but only one billion will have access to drinking water. Twenty thousand children are affected every day due to polluted drinking water which is more than the total mortality caused by cancer, AIDS, wars and accidents. There is an urgent need to find a solution to safe drinking water availability for our people. Drinking water can be obtained by the Reverse Osmosis process from sea water. Carbon nanotube filters are efficient enough to remove micro-to nano-scale contaminants from water and heavy hydrocarbons from petroleum. Made entirely of carbon nanotubes, the filters are easily manufactured using a novel method for controlling the cylindrical geometry of the structure. The filters are hollow carbon cylinders several centimetres long and one or two centimetres wide with walls that are only one-third to one-half a millimetre thick. They are produced by spraying benzene into a tube-shaped quartz mould and heating the mould to 900°C. The nanotube composition makes the filters strong, reusable, and heat resistant, and they can be cleaned easily for reuse. Carbon nanotube filters offer a level of precision suitable for different applications. They can remove 25 nanometer-sized polio viruses from water as well as larger

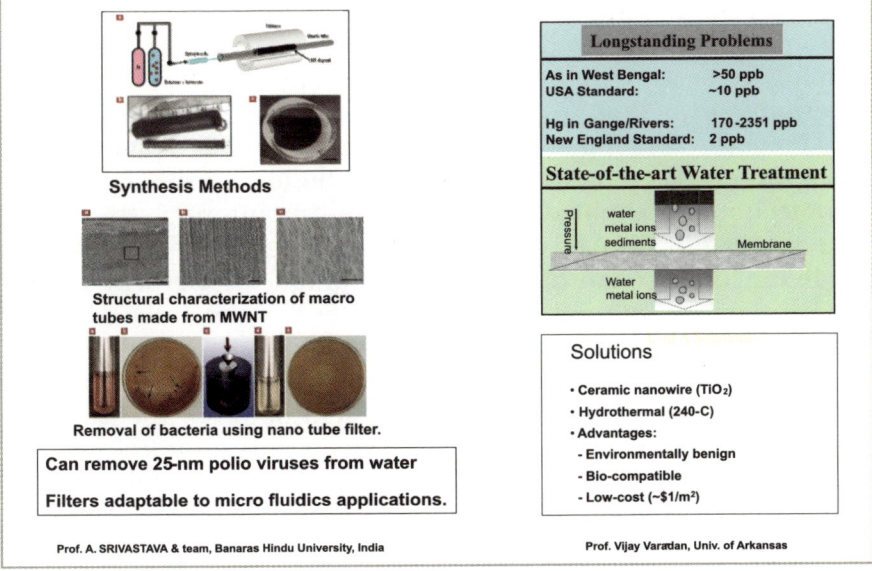

Fig. 3.13: Nanotechnology Application for Water

pathogens such as E.coli and Staphylococcus aurous bacteria (Figure 3.13).

Researchers believe this could make the filters adaptable to micro fluidics applications that separate chemicals. This is a classic application of Nanoscience to the age-old problem of water purification. If properly used, this can help in lessening the burden in our drinking water missions leading to the availability of safe drinking water that will result in minimizing water-borne diseases.

Nanotechnology in Agriculture

Nanotechnology has the capability to revolutionise the agricultural and food industry with new tools for various purposes such as the molecular treatment of diseases, rapid disease detection, enhancing the ability of plants to absorb nutrients, etc. Smart sensors and smart delivery systems will assist the agricultural industry in fighting against viruses and other crop pathogens. Nanostructured catalysts will be available to increase the efficiency of pesticides and herbicides in the near future. Nanotechnology will also protect the environment indirectly through the use of renewable sources of energy, and filters or catalysts to reduce pollution and clean up existing pollutants.

Precision farming has been a long desired goal to maximise crop yields while minimising input (i.e. fertilisers, pesticides, herbicides, etc) through monitoring environmental variables and applying targeted action. Computers, global satellite positioning systems, and remote sensing devices play a key role in implementing precision farming to measure highly localised environmental conditions thus determining whether crops are growing at maximum efficiency or not. Precision farming also contributes to reduction of agricultural waste thereby keeping environmental pollution on the lower side. Nanotechnology enabled tiny sensors and monitoring systems will have a great impact on future precision farming methodologies.

In the future, nanoscale devices with unique properties could be used to make agricultural systems "smart". For instance, devices could be used to identify plant's health issues well before they get noticed. Nanotechnologists are optimistic not only about the potential to change the existing food processing system but also to ensure the safety of food products thereby creating a healthy food culture. Besides, they are also hopeful of enhancing the nutritional quality of food through selected additives and improvements in the way the body digests and absorbs food.

Nanotechnology is already making an impact on the

development of functional foods which respond to the body's requirements and can deliver nutrients more efficiently. A key element in this sector is the development of nanocapsules that can be incorporated into food to deliver nutrients.

The emerging field of nanotechnology for agriculture appears quite promising. However, the potential risks in using nano particles in agriculture cannot be left unnoticed. With the current application and advancements soon to follow, nanotechnology is bound to have a great impact on the direction that agriculture will take. Scientists are laying track for new technologies and looking at every possibility to improve upon current methods. In agriculture, there are still many possibilities to explore and a great deal of potential in upcoming products and technologies.

Emerging Nano Application in Healthcare

The core areas of Bio, Nano and Info technologies combined together will bring about a technology revolution where the effects are going to echo in all the related fields of healthcare transformation. The whole concept will be to provide total solutions in healthcare for all. The important areas which will see the revolution are nanomedicine, nanodiagnostics, DNA repair, genetic modifiers, etc. Some of the frontier areas will witness the wonders of these new miniaturised systems (Figure 3.14). They would be capable of repairing living cells, delivering drugs through nanorobots or tissue culture which could be in the critical human organs as liver, kidneys, etc. Multi-walled carbon nanotubes would be going into the device fabrication of bio and nano based sensors for a large variety of applications from micro syringe to MEMS based implantable medication delivery systems.

Nano bio medical sensors will play a major role in the early detection of dreaded diseases like AIDS, cancer, etc. Nanochips automatically monitor blood sugar levels for diabetics and discharge drugs automatically in appropriate quantity and at the right time. Nanotechnology offers a wide range of opportunities such as synthetic scaffolding and nanoceramic coatings for tissue repair and implants. The present practice of painful electric shocks for the treatment of lunacy will be eliminated and would be replaced with specific localized electric and magnetic stimulation of the brain parts with minimal pain for the patients. Further, bio-nano technology devices such as nano machines (Nano robots) would help in the cleaning of blocked arteries leading to prevention of heart attacks

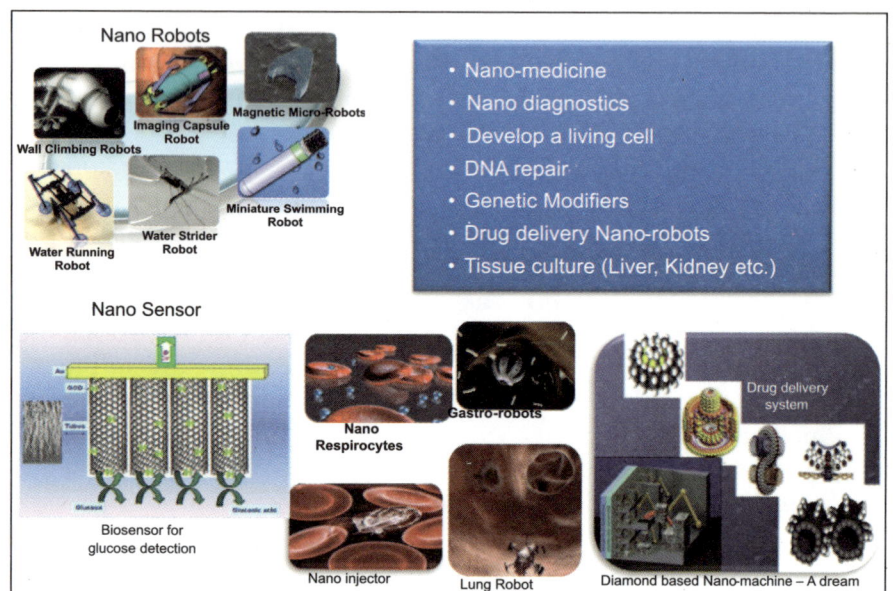

Fig. 3.14: Emerging Areas in Bio-Nano Technology

and destroying the viruses using nano lasers: Nano neuro transmitters would repair neurons in the brain using biochips and would cure diseases like Parkinson's, Alzheimer's and Epilepsy.

Prof. Vijay K Varadan of the University of Arkansas shared his experience of the development of a wireless system to link with the bio-nano chip in the brain for possible treatment of Parkinson's disease, Alzheimer and Epilepsy (Figure 3.15). The primary symptoms in Parkinson's disease are trembling of hands, arms, legs, jaw and face, rigidity or stiffness of the limbs, slowness of movement and impaired balance. The tremor is measured by a passive polymer based sensor on the wrist near the tremor location. The sensor gets power from a pacemaker which also reads tremor motion. The pacemaker then generates a pulse in the implanted device in the brain to activate neurons. Once the neurons are activated, the tremor is controlled.

This is a promising area of research in which we will soon have a breakthrough for the application of carbon nanotube based biochip and the wireless system. We can see the urge of the international scientific community to use nanotechnology based bio devices to cure incurable diseases and make humans function normally. It is only a matter of time that the concept of incurable disease will be redefined.

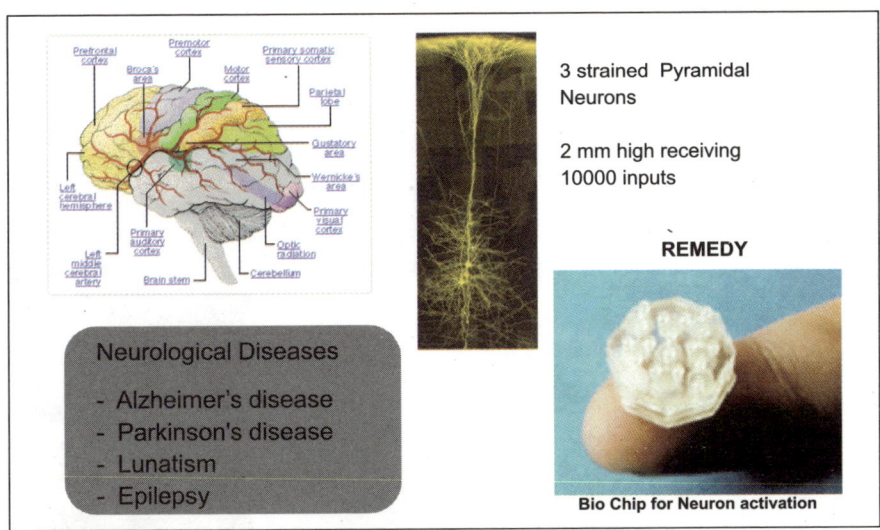

3 strained Pyramidal Neurons

2 mm high receiving 10000 inputs

REMEDY

Neurological Diseases

- Alzheimer's disease
- Parkinson's disease
- Lunatism
- Epilepsy

Bio Chip for Neuron activation

Fig. 3.15: Application of Nano Bio Sensor

Nano Medicines and Nano-therapeutics

Applications of nanotechnology are currently being developed for a broad range of therapeutic applications. It is expected that over the next five years nanotechnology will result in significant, and perhaps paradigm-changing, advances in early detection, molecular imaging, assessment of therapeutic efficacy, targeted and multifunctional therapeutics, pulse laser for nanopharmacy and prevention and control of diseases like Cancer, TB, and others. Specific tissue targeting of nanoparticles for diagnostic purposes or drug delivery is being pursued on a number of fronts (Figure 3.16). Many types of cells and tissues can be distinguished on the basis of extra cellular markers. Homing peptide ligands that attach to nanoparticles and direct them to recognize specific extracellular markers are being developed for a wide range of cell types. The vasculature is an attractive tissue for targeting efforts because malignant tumours must recruit their own blood supply in order to sustain their uninhibited growth. Tumour vasculature can usually be distinguished from normal blood vessels permitting the development of specific, targeted therapeutics to attack and cut off critical blood supplies to tumours.

Other delivery systems employ nanotubes developed by materials scientists and engineers. Reinforcement of nanotubes increases the strength and elasticity of plastics and other materials. These nanotubes are capable of crossing cell membranes and ferrying

Target Specific Delivery System

- *Enable to use lower doses*
- *Could be avoid saturation of whole body with drug*
- *Minimization of side effects*
- *Possible to use stronger doses which could not possible using conventional methods*
- *Example: use of gold coated spheres to heat and destroy the cancerous cells.*

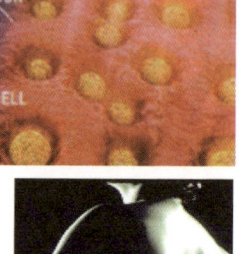
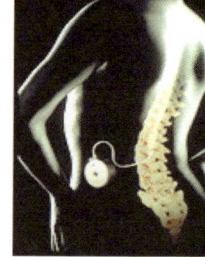

Fig. 3.16: Nano Drug Delivery

Nanotechnology offers tools and techniques for more effective detection, diagnosis and treatment of diseases

<u>Detection and Diagnosis</u>

- **Lab on chips help detection & diagnosis of diseases more efficiently**
- **Nanowire and cantilever lab on chips help in early detection of cancer biomarkers**

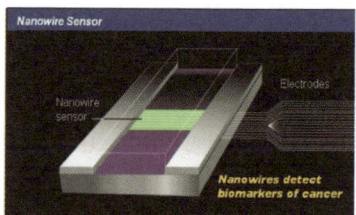

The microfluidic channel with nanowire sensor can detect the presence of altered genes associated with cancer
– J. Heath, Cali. Insti. of Technology

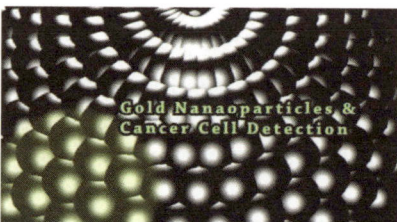

Curing of Cancer using Gold Nanoparticle

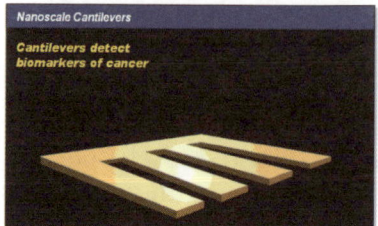

The nanoscale cantilever detects the presence and concentration of various molecular expressions of a cancer cell
– A. Majumdar, Univ. of Cal. at Berkeley

Fig.3.17: Detection of Cancer & Curing of Cancer Cells through Nanoparticles

proteins into cells. Because eliminating specific RNA molecules has been shown to bring down regulate cancer growth, attaching RNA-degrading enzymes or antisense molecules (RNA or DNA that interfere with gene expression) to nanotubes is one approach to under development. In one recent study, complexing antisense DNA with gold nanoparticles was demonstrated to boost delivery into cancer cells. In addition, the antisense molecules also bound to their targets more efficiently.

Nanotechnology based diagnostic and therapeutic tools are also being developed for medical conditions such as cardiac disease and neurological disorders (Figure 3.17). Because they can cross the blood-brain barrier, nanoparticles hold enormous promise for efficient drug delivery systems for brain disorders. In one recent study, an analgesic drug attached to albumin nanoparticles was successfully delivered to the brains of mice.

Nanoparticles find many applications in the cure of the dreaded diseases. Scientists at MIT have designed a new type of nanoparticle that could deliver vaccines effectively for diseases like HIV and Malaria. Synthetic Protein vaccines are covered by fatty droplets called liposomes which increase the T-cell response (Figure 3.18).

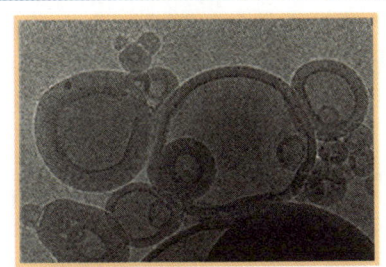

Vaccine - delivering nanoparticles by placing lipid spheres inside one another (Image: Peter DeMuth and James Moon)

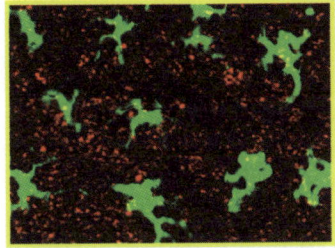

Immune cells, tagged with green fluorescent protein, are surrounded by nanoparticles (red), after the nanoparticles are injected into the skin of a mouse (Image: Peter DeMuth and James Moon)

- New type of nanoparticles consist of concentric fatty spheres that can carry synthetic versions of proteins

- Fatty droplets called liposomes packaged like a virus-like particles in a concentric spheres promote T Cell responses

- On absorption by a cell, nanoparticles degrade quickly and release Vaccine.

- These nanoparticles safely & effectively deliver vaccines for infecticious diseases

At Trial Stage   Source: MIT

Fig.3.18: Vaccine delivering Nanoparticles for HIV & Malaria

Once absorbed by a cell, these droplets degrade quickly and release vaccine provoking T cell response. These are at the trial stage and soon this would be a reality in treating the human beings affected with HIV and Malaria.

Nano Bio medical sensors will play a major role in the early detection of dreaded diseases like AIDS and cancer. Immuno assays can be used for detecting antigens in blood samples by the introduction of nanoshells attached to antibodies while sampling blood. Nanomagnetic particles can be used for targeted treatment and diagnosis. Nanosensor based smart drug delivery systems discharge drugs in appropriate quantity and at the right time. Similarly, nanotechnology offers a wide range of opportunities such as synthetic scaffolding and nanoceramic coatings for tissue repair and implants respectively.

Bio-Nano Technology in Stem Cell Therapies

The convergence of bio-nano stem cells will revolutionize healthcare. Application of nanotechnology in stem cell therapies has great opportunities. Stem cell nanoparticle further enhances pharmacological efficacy. Nanoparticles can act on living cells at the nano level resulting in biologically desirable effects. Magnetic nanoparticles combined with antibody CD34 enrich Peripheral Blood Progenitor Cells (PBPCs) that are more effective than bone marrow transplantation for clinical oncology.

Magnetic nanoparticles have a significant impact on both clinical oncology and basic cancer research. Mesenchymal stem cells have the ability to differentiate into all cell types including neurons, cardiomyocytes, hepatocytes, islet cells, skeletal muscle cells, and endothelial cells. Bioactive materials such as natural matrix components like fibres, and growth factors incorporated into the fibres during fabrication provide an active surface of the scaffold that enhances cellular functions. This type of multifunctional scaffold system is useful in regenerative medicine. Nanowires guide the differentiation of stem cells into specific tissue types through electrical pulse or chemical transmittance. Quantum dots have also been used for tumour targeting and imaging, lymph node and vascular mapping. Quantum dots help in monitoring the survival of cells after transplantation. Nano-topography plays a vital role in regulating stem cell differentiation and provides information about changes in cell shapes and gene expression.

Nano-Robots

The Harvard School of Engineering and Applied Sciences has experimented with the use of nano needles for piercing and delivering the drug content into individual targeted cells. That is how nanoparticle science is shaping bio science. In the same laboratory, DNA materials are made to form self assembling particles. When a particular type of DNA is applied on a particle at atomic level, it will generate a pre-fixed behaviour and automatic assembly from them. This could be the answer to self assembly of devices and colonies in deep space without human intervention as envisioned by Dr. K. Eric Drexler. Thus, in a single research building, two different sciences are shaping each other without any iron curtain between the technologies. Convergence of science is reciprocal. The reciprocal contribution of science is going to shape our future and industries need to be ready for it.

Nanocomputers

Molecular switches and circuits along with nanocell will pave the way for the next generation computers. Ultra dense computer memory coupled with excellent electrical performance will give the society low power, low cost, nanosize and yet faster assemblies. The cross-bar memory with molecular switches and nanowires will drastically reduce the bit size from 80nm to only 8 nm and the data storage from 100 GB/sq. inch to 10 TB / sq. inch.

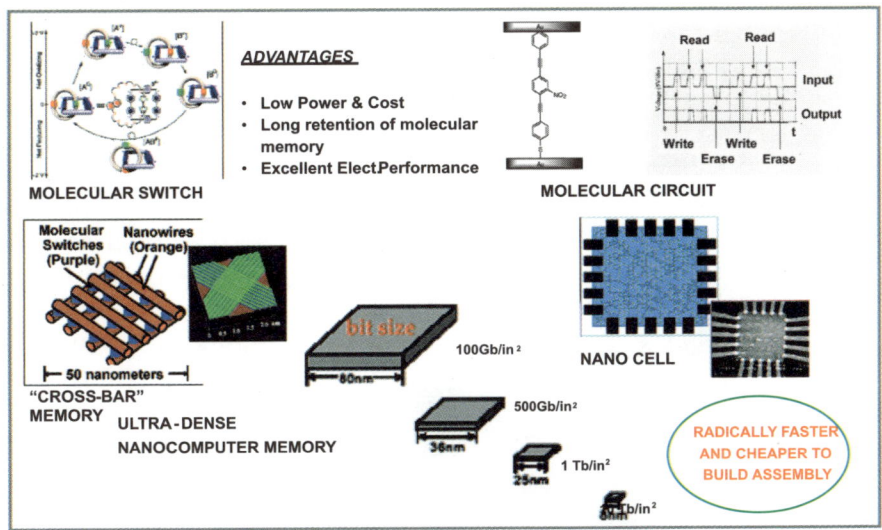

Fig. 3.19: Next Generation Nanocomputers

Nanotechnology Status in India

In India, modest beginnings have been made by some of the institutions which are contributing towards this pioneering research.On October 2001, the Department of Science and Technology launched a major Nano Science and Technology Initiative (NSTI). The broad priority areas included research in nanoscience and technology, strengthening and building infrastructure, generation of trained manpower and having interface between academy and industries, establishing Centres of Excellence and strengthening the characterization facilities, initiating joint institution industry linked projects and Public Private Partnership activities. Some of the important areas where application research is pursued are nanotube, nanolithography and nanoelectronics, drug/gene targeting, DNA Chips, nano-structured high strength materials, quantum structures etc. The current status of research in nanoscience and technology comprises 5613 research papers published under eight broad areas such as Nanocomposite structures, NEMS and MEMS, microelectronics, High energy materials, Stealth and Camouflage, NBC Devices, Modelling and Simulation of nanomaterials, testing and Characterisation. The researches undertaken by the scientists have motivated them to file number of patents. There are near nearly 185 patents that have been

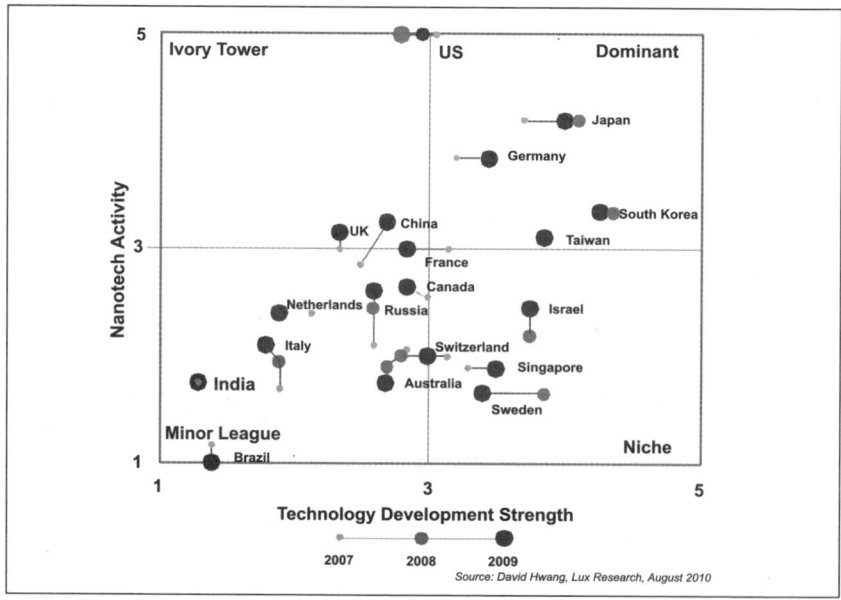

Fig. 3.20: Ranking the Nations on Nanotechnology

granted till October 2011 to scientists, researchers and industrialists based on their affiliations to University, R&D labs, Industries and institutes. Out of this, Industries hold 59 per cent of the granted patents followed by Technical institutions with 17 per cent. Next come Universities with 14 per cent and finally R&D with 10 per cent. It is essential that the research by academia and R&D Labs in achieving the realisation of a particular technology product has to be transferred to the Industry. According to a study undertaken by Indian Academia & R&D Labs by National Foundation of Indian Engineers, so far, twenty one nano technologies developed in the public domain are transferred to the Industry. This shows that the Nanotechnology industries are showing keen interest in the commercialisation of nano products keeping in mind the potential future market of nano products.

According to David Hwang, Lux Research report 2010 on ranking of the nations on nanotechnology development in 19 countries around the world by analysing their performance on the two axes of Nanotechnology activity and Technology development (Figure 3.20) shows USA placed at the top of the graph with high nanotech activity but below average in the Technology Development strength. Japan has the second highest nanotech activity with good technology development strength. China is growing steadily. Russia is making a great push through its State-owned RUSNANO but still among the minor league countries. India though its NSTI mission has started its research activities in nanotechnology and commercialised some of the technologies, still remains in the minor league category where both nanotech activity and technology development strength are inconsequential to the development and commercialisation of nanotechnology. India should give more thrust to nanotechnology research and the commercialisation of nano products thereby increasing the technological strength in the nanotechnology activity.

Business Impact of Nanotechnology

The business impact of nanotechnology is going to be on a higher side in the years to come. Many nanotechnology products have started coming and according to a survey, 50 per cent of the new products in advanced industrial areas will use nanoscience and engineering by the year 2015. The revenue impact of nanotechnology will be on a lower side for machinery, food and paper whereas motor vehicles, apparels and textiles are going to have

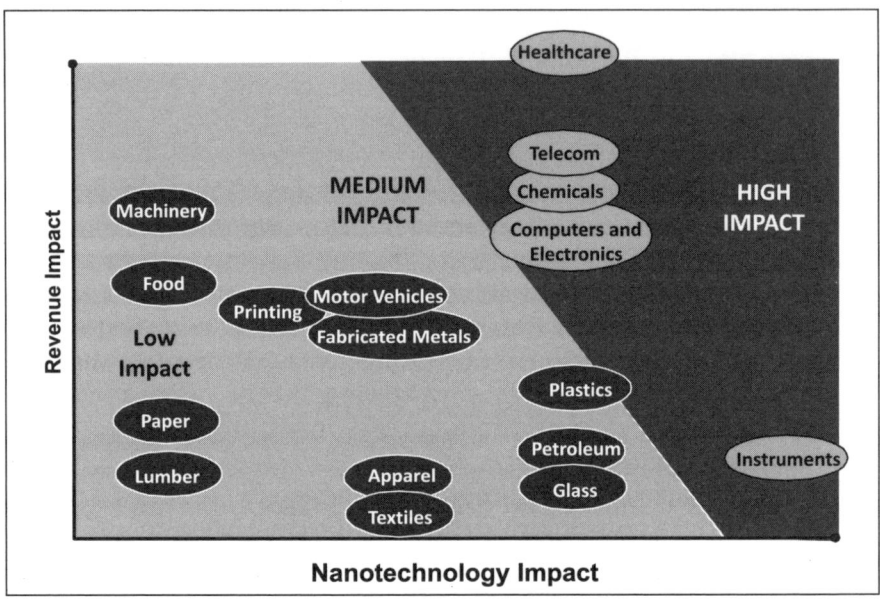

Fig.3.21:Nanotechnology Business Impact

medium impact. High impact zone areas are instruments, computers and electronics, chemicals and to top them all it is the Healthcare sector (Figure 3.21). During the 2000s the focus was on the fundamental R&D discoveries in nanotechnology. Gradually, the market for final products incorporating nanotechnology gained momentum and stood at $254 billion in the year 2010. According to the Nano2 Report, 2010, the market for nanosystems in the year 2020 is predicted at $3 trillion.

Every product of the future will have an element of nanotechnology in it. Therefore, it is essential that major thrust is given to formulate the development effort with clear-cut missions and achievable results for many applications in military and civil sectors. Partnership with universities and industries will therefore synergise all our efforts to benefit the country. Nanotechnology encompasses all branches of Science and Technology at the molecular level and the developments are to be in line with nature acceptable principles and safety. The curriculum designers of technological courses may like to take this aspect into account while formulating new courses. India missed the great industrial revolution and became only a developing country. Now it has the potential in the knowledge age to excel in research in this advanced technology. India should not miss this great opportunity of going closer to the world technology, offered by Nano revolution to regain its glory. Convergent technology is a significant contributor for mankind and it will also be a Game Changer.

3.2.4 Convergence of Technologies

Information technology and communication technology have already converged leading to Information and Communication Technology (ICT). Information Technology combined with biotechnology has led to bioinformatics. Similarly, Photonics is grown out from the labs to converge with classical Electronics and Microelectronics to bring in new high speed options in consumer products. Flexible and unbreakable displays using a thin layer of film on transparent polymers have emerged as new symbols of entertainment and media tools. Nanotechnology is the field of the future that will replace microelectronics and lend many other fields with tremendous application potential in the areas of medicine, electronics and material science.

When Nano technology and ICT meet, integrated silicon electronics, photonics are born and it can be said that material convergence will happen. With material convergence and biotechnology linked to each other, a new science called Intelligent Bioscience will be born which would lead to a disease free and more intelligent human habitat with longevity and high human capabilities. Convergence of bio-nano-info technologies can lead to

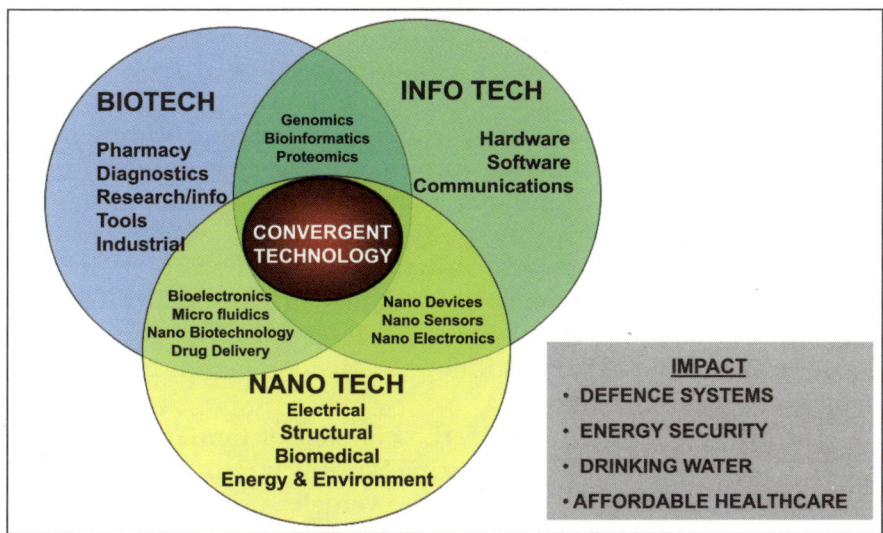

Fig. 3.22: Convergence of Technologies

the development of nano devices like nano robots (Figure 3.22). When Nano robots are injected into a patient, they will diagnose and deliver the treatment exclusively in the affected area and then the nanorobot will get digested as it is a DNA based product. Drug delivery system will revolutionize healthcare to a large extent. Dip pen nanolithography is the latest nanofabrication method which uses microscopic tips coated with a material which can be metals, DNA, or a protein that is deposited as 'ink' on a wide range of surfaces such as silicon, glass or metal. Convergence of technologies opens up new areas for research and ultimately results in the betterment of human beings.

Now a new trend is emerging. Globally, the demand is shifting towards development of sustainable systems which are technologically superior. This is the new dimension of the 21st century Knowledge Society, where science and environment go together. Thus the new age model would be four dimensional Bio-Nano-Info-Eco based.

3.3 ROBOTICS, ARTIFICIAL INTELLIGENCE AND COGNITIVE SCIENCES

3.3.1 Robotics Engineering

The term ROBOT popularised by the Czech Writer Karel Capek in 1921 through his play R.U.R.(Rossum's Universal Robots) from the Czech word "Robota" means hard labour and servitude; the robots have continuously changed in physical appearance and artificial intelligence and today over 1 million household robots and 1.1 million industrial robots are being operated worldwide performing tasks accurately. The emerging robot is a machine with sensors, processors and effectors, able to perceive the environment, make appropriate decisions, and act upon the environment. Various sensors involved are active and passive optical, ladar vision, acoustic, ultrasonic, RF, microwave, touch, etc. Depending on the applications, the robots may be designed as wheeled, tracked or legged, with various control system architectures. Robots use fibre optic cable, RF, laser, acoustic for various command, control and communications systems.

During World War I and II, the concept of unmanned air and ground vehicle developed and during Word War II, the Germans had developed inexpensive tele-robotic vehicle using Electric Dog technology for mine clearing. Later, a tracked vehicle was developed that carried explosive container hooked to the carrier. During the Persian Gulf Wars, development and acquisition of Unmanned Ground Vehicle (UGVs) took place. One area where even more advances in autonomy have been made is the development of unmanned aerial vehicles, or UAVs. These are essentially remote-controlled spy vehicles that are capable of flying themselves if they lose contact with their pilot. These planes can also be used to monitor forest fires. These UAVs had become more popular because of their unmatched surveillance, reconnaissance and the effective delivery of payload without engaging humans.

DRDO has succeeded in realising many robotic systems (Figure 3.23). Versatile remote-operated vehicle DAKSH is used for identification and handling of Improvised Explosive Device. The same can also be used to survey and monitor nuclear and chemical contamination levels. Unmanned airborne vehicle "Vihanga Netra" has been developed for monitoring snow cover and avalanches in remote avalanche terrain. Multi-mission Unmanned Aerial Vehicle NISHANT is used for reconnaissance, surveillance and target tracking with day and night capability. A variety of controllers and manipulators have been developed for Gantry, SCARA and other types of robots. A prototype of Unmanned Ground Vehicle has been developed to attain more autonomous capability. Robots have been developed for Non-destructive testing, Ammunition loading and hot slug manipulation.

Fig. 3.23: Robotic Systems Developed by DRDO

Homeland Security

Robots with the help of artificial intelligence are used extensively in the area of homeland security mainly for surveillance purposes. The bigger size UAV of the United States, Predator, was used for the continuous surveillance of anti social elements by homing on to them and effectively delivered the payload and destroyed them. The drone operators were sitting and operating from their cabin. The advanced high resolution cameras with real time communication are made to execute precisely this type of operation and do not allow the enemy to know and react. The advanced technologies like nanotechnology have enabled the miniaturisation of the robots for special applications like the Nano Air Vehicle or Humming bird of DARPA for surveillance (Figure 3.24). This tiny vehicle resembles a hummingbird and weighs 19 gm equipped with batteries, motor and powerful video camera as a payload. It can fly at 17 km per hour and can move in three axis of motion. This robot could be deployed for reconnaissance and surveillance in urban operations or on battlefields to survey the surroundings.

Robots are being extensively used in space explorations aiming to replace humans from facing dangerous situations that could arise. Robonaut is a robot operated by a remote, designed to perform dangerous space walks in place of an astronaut. In addition, robotic rovers had already been sent to Mars for exploration purposes. Space probes such as Huygens (which landed on Titan) and Russia's Venera 9 (which landed on Venus) are sometimes considered robots too. Not only in space but also in subsea operations, autonomous unmanned

Artificial hummingbird

❖ Wingspan : 16 cm;
❖ Weight: 19 gms
   (less than an AA battery)
❖ Speed: 17 km/h (Three axes)

Source: DARPA

❖ **Contains Nano batteries, motors, & communications systems; as well as the video camera payload**

❖ **Can climb and descend vertically; fly in all directions**

❖ **Manoeuvres using its flapping wings for propulsion and attitude control**

❖ **Could be deployed to perform reconnaissance and surveillance in urban environments or on battlefields**

Fig. 3.24: Nano Air Vehicle

underwater vehicles will have a vital role to play.

Future wars will be fought by unmanned entities rather than soldiers. Replacement of humans by machines to operate in hazardous areas and in a combat scenario is the thrust area of research in military technology and is being pursued worldwide. The choice of system level architecture, configuration, sensors and components through conscientious design and simulation provides for significant synergy within a robotic system. Well-designed robotics systems will become self-reliant, adaptable and fault tolerant, thereby increasing the level of achievable autonomy. The enabling technologies of robotics will see increased adoption in military roles. Future research will continue to focus on the perception and control of robotics vehicles in an effort to increase the baseline level of autonomy and utility for military applications. Also, the future Robots would have better actuators which would ease their movement and better sensors to know about the environment thus making them smarter. They would also have an electronic skin that can sense delicate and fragile objects. The advent of nanotechnology would open up many new frontiers in this technology.

3.3.2 Artificial Intelligence (AI) and Expert Systems in Agriculture

Application of AI Expert System in Agriculture

A dynamic information system and solutions to problems
- Study of Plant Pathology, Entomology, Soil condition into a framework for assessing farmer's needs
- Weather and climate monitoring
- Irrigation scheduling
- Fertilizer scheduling
- Diagnosis of disorders and treatment
- Crop production and assessment
- Overall assessment of the farm
- Communication to the farmer in local language

Robotic Technology for Agriculture

Field Inspector

The API platform's third generation prototype is a four wheel-drive vehicle, has four-wheel steering with two motors per wheel, one for providing propulsion and the other for steering to achieve higher mobility. The platform has a ground clearance of 0.6m and track width of 1 m. It is operational with Real Time Kinematic Global

- ❖ **Autonomous system requires an Autonomous Plant Inspection (API) vehicle and cameras for weed detection and mapping**

- ❖ **Vehicle has a height clearance of 0.6 m and track width of 1 m**

- ❖ **Equipped with a real time kinematics Global Positioning System (RTK-GPS)**

- ❖ **Operating console on the top frame for implementing agricultural operation (spraying, weeding etc.)**

GPS APPLICATIONS

- Field preparation, Planting and Cultivation
- Fertilizing &Crop Protection
- Mapping, Scouting & Sampling
- Harvesting
- Planning and Analysis

Fig. 3.25: Field Inspector

- ❖ **Robotic weeding is a novel weeding technology to reduce the amount of energy used to weed organic crops.**

- ❖ **Weeding operations are**
  - between the rows (inter-row),
  - within the rows (intra-row); and
  - close-to-crop

- ❖ **Robotic weeding**
  - Tillage for intra-row and
  - Micro spray close-to-crop

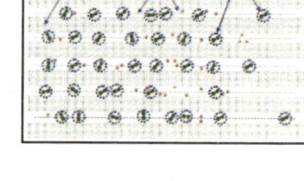
Inter-row   Intra-row   Close-to-crop

Tillage for Inter- row weeding

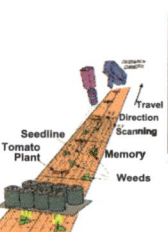
Travel Direction
Scanning
Seedline
Tomato Plant
Memory
Weeds

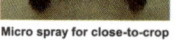
Micro spray for close-to-crop

Source: Danish Institute of Agricultural Sciences

Fig. 3.26: Robotic Weeding

Positioning System (RTK-GPS). An operating console is fitted on the top of the frame for implementation of agricultural operation like spraying devices, sensors or weeding tools. Navigation of predefined route plan and collision avoidance is dealt with by the vehicle's communication with the farm management PC. Possible weed map of the entire field can be created based on the shape recognition of the weeds.

GPS Applications
- Field preparation, Planting and Cultivation
- Fertilising & Crop Protection

- Mapping, Scouting & Sampling
- Harvesting
- Planning and Analysis

Robotic Weeding

There are several methods for killing and removing unwanted plants based on the position and severity of the weeds. Breaking up of the soil and root interface by tillage and promoting wilting of the weed plants could be one of the possibilities. This can be achieved in the inter row area easily by using classical spring or duck foot tines.

Intra row weeding, on the other hand requires the position of the crop plant to be known so that the end effectors can be steered away. Within the close-to-crop area, tillage is not advisable as any commotion to the soil is likely to damage the interface between the crop and the soil.

Laser treatments and micro-spraying are two non contact methods that are being developed. Non-competitive weeds can be left to grow if they are at a handy distance from the crop. This is termed as controlled biodiversity which could be realised using robotic weeding.

Automation in Water Management and Irrigation

The modern automation systems comprises five basic components:
1. Measuring and sensing equipment,
2. Control and regulation instrumentation,
3. Input and output devices,
4. Communication between the different components; and
5. Power sources

Robotic Irrigator

A robotic irrigator in the form of a mechatronic sprinkler that is capable of simulating a travelling rain gun was developed to apply variable rates of water and chemigation to selected areas. Based on the present weather conditions and the preferred pattern a small computer adjusts the trajectory and sector angles of the jet controlled by stepper motors. When the airborne water was blown down the wind, the jet angles could be adjusted to compensate by measuring the instantaneous wind speed and direction. This unique system not only allows the required amount of water to reach at the right place but also irrigates the field corners which generally remain left out.

Implementation of Automation
- Time-based automatic opening and shutdown of the water so that only specific amount of water reaches every corner of the field.
- The water delivering system should open by the help of timer and at the same time should be able to automatically shutdown after required water is delivered.
- Combined irrigation and fertilization with or without recording of the applied water and fertilizer amounts.
- Sequential activation of valves in the field is necessary for restricting wastage of water as well as resisting damage of crop due to over-supply of water.
- Integrated scheduling and control of irrigation systems. Real time control through information received from sensors (Temp., wind, rain, soil moisture, etc.)
- Integrated control of water sources and irrigation systems.

3.3.3 Cognitive Sciences

How does memory work? How do we understand language and produce it so that others can understand? How do we perceive our environment? How do we infer from patterns of light or sound the presence of objects in our environment, and their properties? How do we reason and solve problems? How do we think? These are some of the foundational questions that cognitive psychology examines. The term "cognition" refers to all processes by which the sensory input is transformed, reduced, elaborated, stored, recovered, and used. It is concerned with these processes even when they operate in the absence of relevant stimulation, as in images and hallucinations.

Cognitive Science and Related Terms

Cognitive Science is the scientific study of the mind and behavior of the human being to achieve optimum performance in a given environment. It is the study of relationships among the integration of cognitive psychology, biology, computer science, linguistics, philosophy and mathematics. Cognitive Neuroscience establishes a link between brain functions and behavior. It states that transformation in cognitive process has a neural basis and the process has its locus at the higher centres in the brain. Cognitive Psychology is a study of mental operations that support people's acquisition and use of knowledge. Biological Psychology or Biopsychology is the application of the principles of biology to the study of mental processes and behavior. Cognitive Neuro psychology is a branch of neuro psychology that aims to understand how the structure and function of the brain relates to specific psychological processes.

Assumptions of Cognitive Psychology

There are assumptions that cognitive psychology follows (Von Eckardt, 1993, pp 54). These assumptions are:

- Cognitive capacities of abilities can be studied in isolation (e.g. numerical abilities can be studied in isolation with attention or memory).

- Behaviour of a particular individual is the focus of study and is to be studied in his natural environment.

- Cognitive abilities are generally conceptualized to be independent (autonomous) from non –cognitive abilities (e.g. affect, motivation etc).

- A distinction can be made between normal and non-normal (abnormal) cognition.

- Any research question can be stated in a way to fit an information processing model (i.e. information obtained from any observation can be explained by using the information processing model).

Constituents of Cognition

Cognition refers to the mental processes required in acquiring knowledge through the external or internal world. These processes include sensation, perception, attention, memory, learning, problem solving, producing and understanding language, and making decisions. Sensation refers to accepting the information in the form of energy through external or internal world via the medium of senses. These energies are transformed and interpreted in defining the characteristics of the information in the process of perception. The world is full of sensory information but the individual focuses only upon some of the information about description through the process of attention. Attended stimulus information is stored at different levels of memory, i.e. Sensory, Short Term Memory and Long Term Memory. Stored information is helpful in learning new things and continuous learning helps the individual to solve problems.

Cognitive Science: Models

Connectionist Approach: Cognitive psychologists have also sought to understand the mind's representational and computational qualities via an alternative framework, known as connectionism. Connectionist models typically draw their inspiration from some of the known characteristics of the brain. So, for example, we know that neurons are highly interconnected. Seemingly they can pass information on to neurons with which they are connected, either through inhibiting or enhancing the activity of those neurons. They appear to be able to process information in parallel – neurons are capable of firing concurrently. Besides this there are many more properties. Connectionism describes attempts to build models of cognition out of building blocks that preserve these important properties of neural information processing. Typically, researchers simulate connectionist networks on a computer, networks that involve a number of layers of neuron-like computing units. The appeal of connectionism lies in the hope that connectionist models may ultimately stand a better chance of being successful models of cognition.

Cognitive Neuroscience Model: It deals with structure or signal of the nervous system for explaining cognitive functions. Cognitive neuroscience is a combination of both psychology and neuroscience, and also associated with disciplines such as physiological psychology, cognitive psychology and neuropsychology. Cognitive neuroscience relies upon theories of cognitive science coupled with evidence from neuropsychology, and computational modelling.

Neuroscientific Methods to Study Cognitive Processes

The major aim of neuroscientific methods is to understand the underlying brain process for the desired behavior. Three major issues are to be taken care in understanding the brain process, i.e. Lateralization, Regionalization and Localization. Lateralization refers to the hemispheric dominance of brain, whereas regionalization shows involvement of brain region in a particular behavior. When a specific location of brain structure is responsible for performing the activity, the localized structure is referred to as localization of activity. These issues are dealt within the experimental and clinical paradigm in exploring the brain behavior relationship. Following are some methods to study brain behaviour relationship.

Deficits and Lesions

It is one of the oldest methodologies to study the brain function. In a damaged brain, a patient's functioning brain is analyzed in the light of the resulting deficit after the lesion for that specific region of the brain. The goal of such study is to identify the psychological deficits associated with the damaged region. Resulting loss in the psychological functioning is contributed to the lesioned brain area and its normal function is generalized for the normal population. One such classical example of brain damage study was conducted by Paul Broca, and his pioneer work resulted in the well known inferences drawn for localization of language ability.

Single Cell Recording

One way to study the brain processes is recording the electrical activity as the brain performs different functions. Two different methods of recording the electrical activity have been widely explored in the field of cognitive neuro-psychologyThe first method can be employed by placing the electrode on scalp to pick up aggregate electrical currents. This technique includes electroencephalogram (EEG). Another way to record the electrical active of neuron is through the individual neurons, either by inserting electrode inside the neuron or by placing it on the neuron's vicinity. Such methods to study the brain activity suffer from the feeble strength of the voltage and it therefore needs signal amplification.

Neuroimaging

This is the most recent technique in the cognitive neuroscience and has played a significant role in the development of cognitive neuroscience. The neuroimaging techniques revealed similar results that were obtained from a single cell recording activity of the brain. However, neuroimaging is a noninvasive technique and can be used for studying the brain mechanism of higher order cognitive function in normal humans. This technique also made it possible to implement the experimental design involving simultaneous recording of brain activity while carrying out the cognitive task of abstract reasoning, making judgment or perceiving a visual stimuli etc. There are two neuroimaging methods to study the brain i.e. structural and functional neuroimaging. MRI and CT Scans are based upon the structural imaging methods while fMRI, PET, SPECT etc are functional methods. Neuroimaging methods follow BOLD (Blood Oxygen Level Dependent) technique while studying the brain processes.

Cognitive Science Domains

Cognitive Style: The way individuals think, perceive and remember information. Understanding cognitive style facilitates is understanding propagandas, rumors and orientation to complex environmental operations. Therefore research can include areas of Perceptual detection and deception, route finding difficulties, and relative position of objects.

Cognitive Decision-Making: Deciding on important matters and the process of making choices or reaching a conclusion from two or more alternatives. Studies can be undertaken in the areas of decision making under severe stress/time pressure, neural basis of decision under risk and ambiguity, Neurobehavioral correlates of controlled and automatic decision making, and cognitive load in decision making for time and space critical performance.

Meta Cognition: Knowledge of your own thoughts and the factors that influence thinking. Meta cognition helps in understanding strategic decision making, strategic planners and synthesizing information overload. Areas of research under meta cognition include task analysis (based on experiences and knowledge), shared mental models and operational effectiveness.

Cognitive Profiling: Display performance of several different kinds of cognitive tasks/ or Cognitive Profile Inventory is designed to identify the style of thinking, learning and making decisions. Research areas include assessing cognitive abilities in cockpit, military force within large-scale training simulations and cognitive abilities in different cultural contexts (sensory discrimination, perceptual constancy, memory, language in verbal and non verbal communication).

Cognitive Regulation: Ability to cognitively regulate emotional responses to aversive events. Difficulty in sequential and parallel processing, planning and impulse control, visual Perception, and cognitive flexibility are few areas which can be further explored.

Cognitive Multi-Skilling: Training of soldiers to do a large variety of tasks at one workplace. It helps to understand versatility of operation management in modern warfare, simulator based training, recruitment and basic training in trade allocations. Research can be done in the areas of cognitive demands and utilization of sensory modalities during multi-task performance, ability to adapt to new cultural settings (Involve multiple facets, cognition, motivation, behavioral features), multi-tasking & task switching.

Cognitive Retraining: It is the process of changing sub conscious thoughts. Areas of research include attention, concentration, impulsivity, memory training (recall & organization of information), Reasoning and problem solving (solve: specify, option, listen, vary & evaluate), visual perception (tracking, organization, sequencing etc) and decision making (logical, intuitive, creative decisions)

Cultural Cognition: The tendency of individuals to confirm their beliefs about disputed matters of fact to values that define their cultural identities. It consists of cultural adaptation, cultural cognitive style, shaping of cultural values by taking the risk of new technologies and cognition involved in understanding a culture.

With the advancements in biotechnology, robots and artificial intelligence leading to many autonomous systems, cognitive science plays a very important role in making human resources to the best of the ability.

3.4 SENSOR TECHNOLOGY

The future war theatre will have Chemical, Biological, Radiological and Nuclear (CBRN) contamination, as we see globally many countries are building up large infrastructure and preparing for war. Advancements in materials science and engineering have been important drivers in the development of sensors required for the above warfare. This development has led to advanced sensors which are required for identifying the type of chemical or biological agent or radiation in laboratories and in fields.

There are many other types of sensors which are used for industrial applications, defence and counter terrorism. The high resonance stability of single-crystal quartz, as well as its piezoelectric properties has made extraordinary range of high performance sensors which are affordable and have played an important role in everyday life and national defence. More recently, a new era in sensor technology was ushered in by the development of large-scale silicon processing, permitting the exploitation of silicon to create new methods for transducing physical phenomena into electrical output that can be readily processed by a computer. Development is in

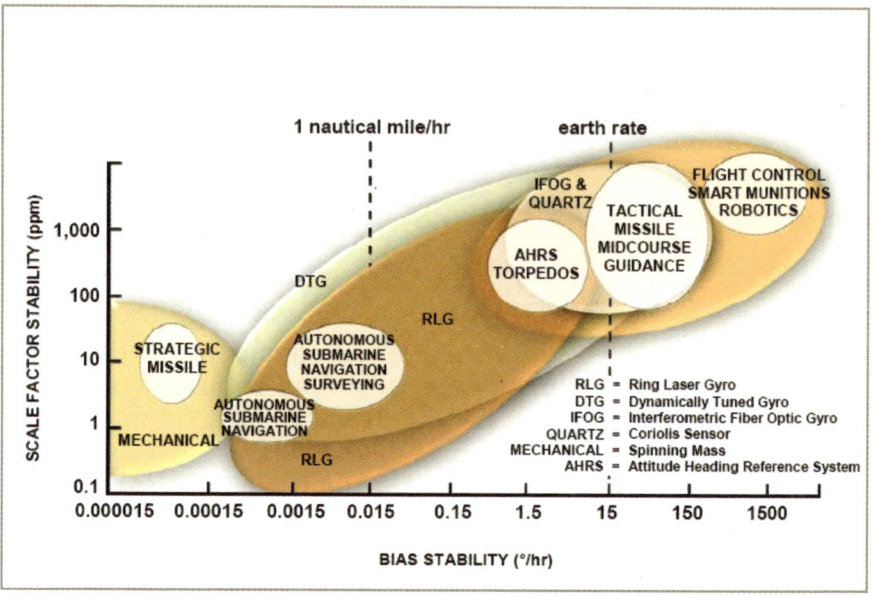

Fig 3.27: Gyro Technology Applications

(Image: George T Schmidt, RTO-EN-SET-064)

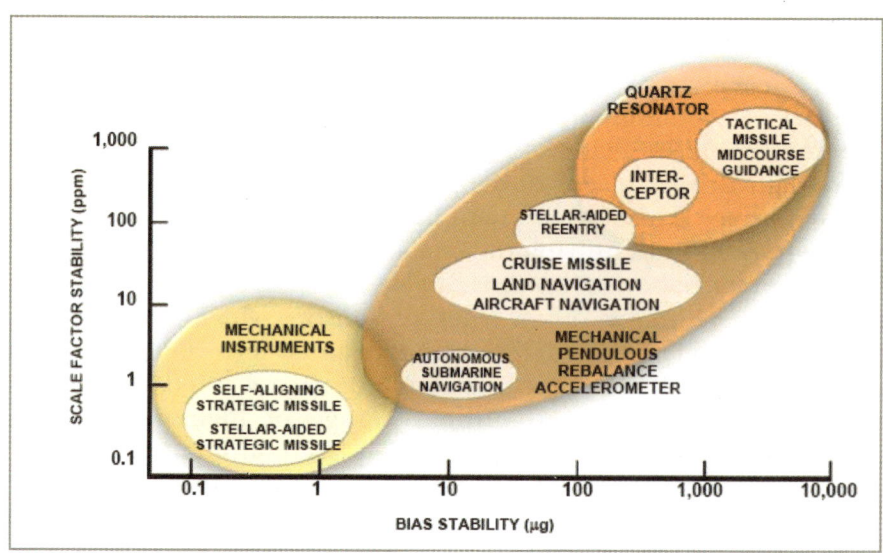

Fig. 3.28: Accelerometer Technology Applications

(Image: George T Schmidt, RTO-EN-SET-064)

progress for better control of material properties and behaviour, thereby offering possibilities for new sensors with advanced features such as greater fidelity, lower cost, and increased reliability. Some of the specific sensors are brought out in the following paragraphs.

3.4.1 Inertial Sensors – Gyros and Accelerometers

Inertial sensors are those that can sense inertial parameters for navigation by determining the position, velocity and acceleration, irrespective of the frame they are in. Gyroscopes and accelerometers are the inertial sensors which are used in launch vehicles, satellites, missiles, aircraft, submarines, ground vehicles and platforms. The application of different types of gyros, namely, Dry Tuned Gyro (DTG), rate integrating gyros in stabilised platforms, Ring Laser Gyros (RLG), Fibre Optic Gyros (FOG), Vibrating Gyro etc. is explained in Figure 3.27 and accelerometers in Figure 3.28. Combining Gyros and accelerometers with electronics, navigation systems are realised for different applications. These navigation systems called INS when integrated with GPS and GLONASS, achieve greater accuracy in flight missions. The breakthrough in Indian guidance system has been the achievement of greater accuracies by software integrations and sensor fusion. RLG based INS + GPS + GLONASS systems have been realised for canister launched missiles like BRAHMOS with fast reaction time and also high accuracies at the target point.

3.4.2 Global Positioning System (GPS)

Constellation of satellite based network to provide location, navigation, tracking, mapping and timing is called Global Positioning System. NAVSTAR GPS satellite system with 24 satellites orbiting the earth at 20,000 km making two complete orbits in less than 24 hours is the American based satellite system providing global navigation coverage. These satellites are arranged in 6 orbital planes with 4 satellites per plane as in Figure 3.29. A worldwide ground control and monitoring network provide the health and status of the satellite and also uploads navigation and other data to satellites.

Global Navigation Satellite System

Fig. 3.29:
GPS Satellite System

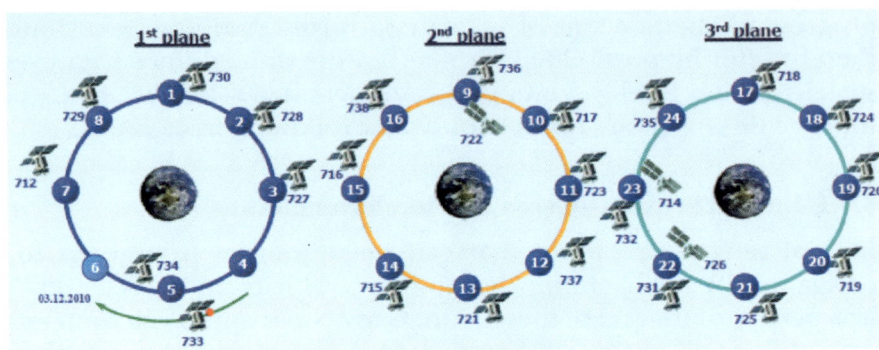

Fig. 3.30: GLONASS

GLONASS is the Russian counterpart to NAVSTAR GPS. GLONASS envisaged the creation of a constantly operating orbit constellation at an altitude of 20000 km, comprising 24 satellites as in Figure 3.30. This consists of a constellation of satellites in Medium Earth Orbit, a ground control segment, and user equipment. This could provide data relating to positioning, direction and speed of a moving object.

The European Union is using Block IIR-M Satellite and is planning to go for full satellite navigation system called Galileo Satellite System with 30 satellites positioned in 3 circular medium earth orbit planes at 23222 km altitude. Chinese BeiDou System, which is now semi-operational, will have ultimately 30 satellites in geostationary orbit. India has plans to have GAGAN, a satellite based augmentation system based on GPS to provide satellite based navigation services for civilian and aviation applications over Indian air space (Figure 3.31) with Indian Regional Navigation Satellite System (IRNSS) (Figure 3.32). This consists of seven satellites, three in geostationary orbit and four in geosynchronous orbit. The full constellation is expected to be operational in 2015.

3.4.3 Sonars

Sonar stands for SOund Navigation And Ranging technique that uses sound propagation usually in underwater to navigate, communicate with or detect objects on or under the surface of the water such as other ships. There are two types of sonar namely 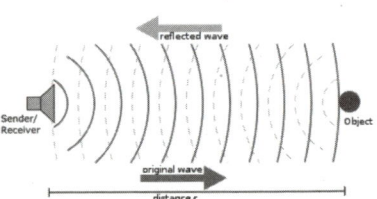 passive sonar and active sonar. Passive sonar is essentially listening for

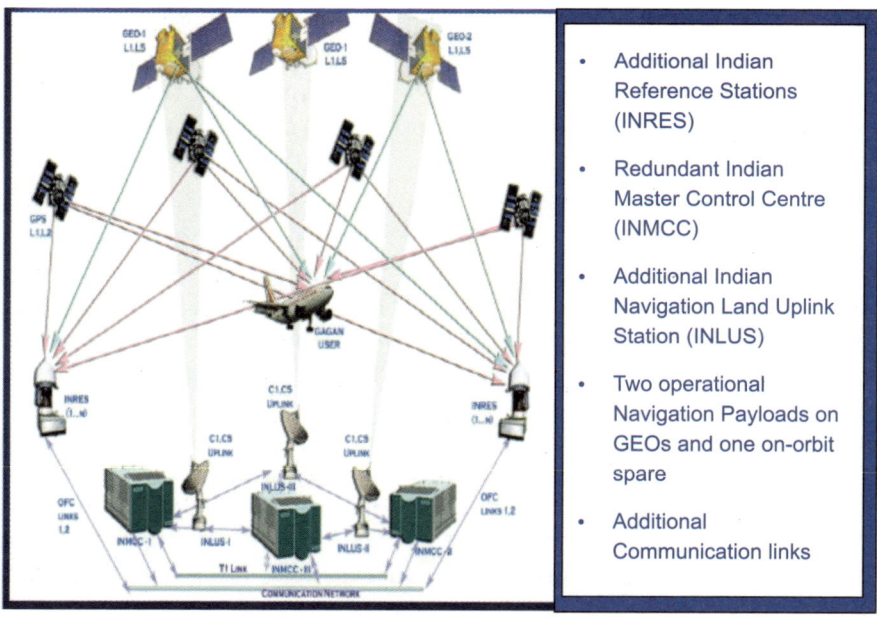

Fig. 3.31: GAGAN FOP Configuration

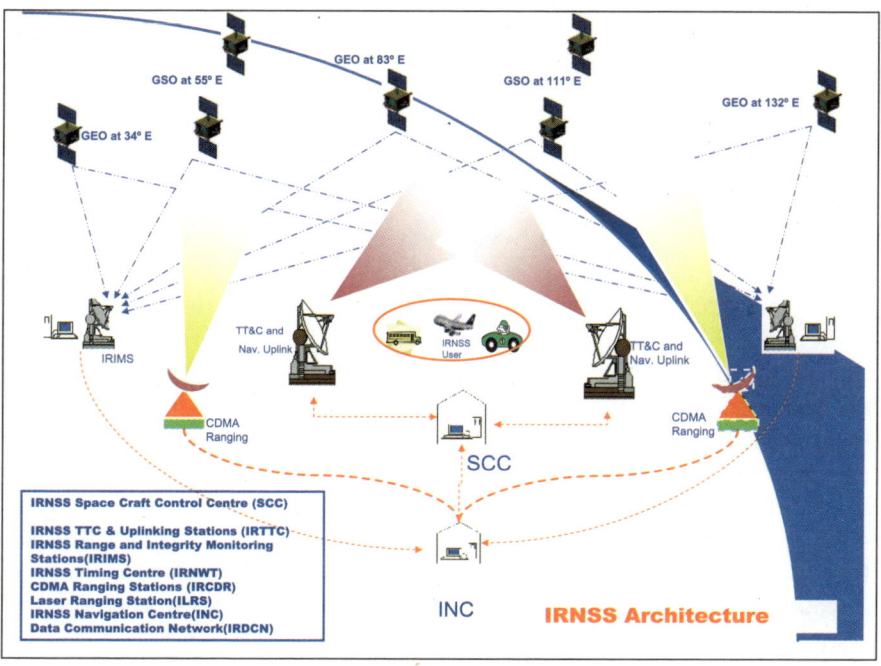

Fig. 3.32: IRNSS Configuration

the sound made by vessels and active sonar is emitting pulses of sounds and listening for echoes. The acoustic frequencies used in sonar systems vary from very low (infrasonic) to extremely high (ultrasonic). The study of underwater sound is known as underwater acoustics or hydro-acoustics.

Active Sonars: Active sonar creates a pulse of sound, often called a "ping", and then listens for an echo of the pulse. This pulse of sound is generally created electronically using a sonar projector consisting of signal generator, power amplifier and electro-acoustic transducer/array. A beam former is usually employed to concentrate the acoustic power into a beam which may be swept to cover the required search angles. Generally, the electro-acoustic transducers are of the Tonpilz type and their design may be optimised to achieve maximum efficiency over the widest bandwidth in order to optimise performance of the overall system. Occasionally, the acoustic pulse may be created by other means, e.g. (1) chemically using explosives, or (2) airguns or (3) plasma sound sources.

Passive sonars: They are often employed in military setting, although they are also used in science applications, e.g., detecting fish for presence/absence in the study of various aquatic environments (see also passive acoustics and passive radar). In the very broadest usage, this term can encompass virtually any analytical technique involving sound generated by remote, though it is usually restricted to techniques applied in an aquatic environment.

Naval Warfare

Modern naval warfare makes extensive use of both passive and active sonar from water borne vessels like ship & submarine, aircraft and fixed installations. India has embarked on the development of underwater sensors to make the country self-reliant in the development of different types of sonars (hull mounted, submarine, air borne and towed array). These sonars are supported by state-of-the-art signal processing techniques and introduction of micro electronics for miniature devices. The indigenously developed sonars have been installed in many of the naval ships. The DRDO developed sonars at Naval Physical and Oceanographic Laboratory (NPOL), Kochi and produced at BEL, are at par with the existing sonars of the developed countries. The growth profile of sonar technologies and systems in DRDO is given in Figure 3.33.

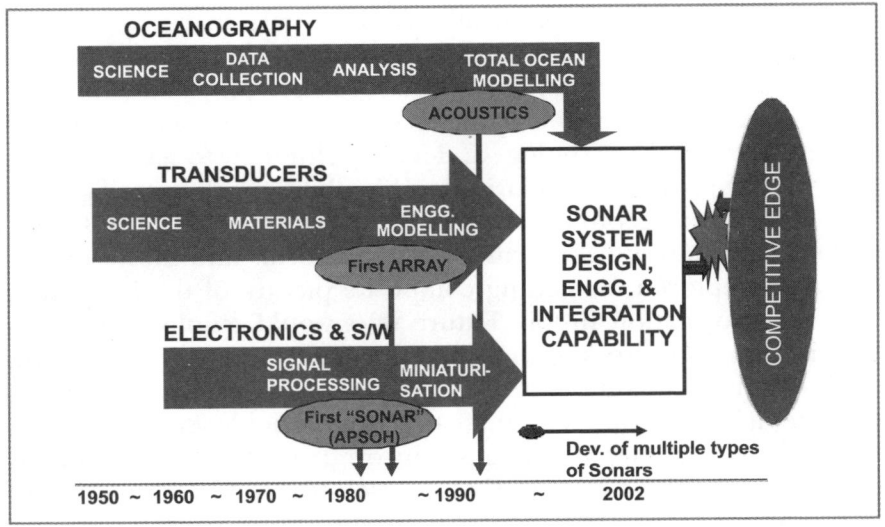

Fig. 3.33: Growth Profile of Sonar Technologies and Systems in DRDO

3.4.4 MEMS

Micro-Electro-Mechanical Systems (MEMS) is a technology that in its most general form can be defined as miniaturized mechanical and electro-mechanical elements such as devices and structures that are made using micro-fabrication. Components of MEMS are micro-sensors, micro-actuators, micro-electronics and micro-structures. MEMS sensors include sensing temperature, pressure, inertial forces, chemical species, magnetic fields, radiation, etc. Remarkably many of these micro-machined sensors have demonstrated performances exceeding those of their macro-scale counterparts. That is the macro-machined version of, for example, a pressure transducer usually outperforms a pressure sensor made using the most precise micro-scale level machining techniques. For instance, Micro-actuators to control gas and liquid flows, optical switches and mirrors to redirect or modulate light beams, micro-mirror arrays for displays, micro resonators for a number of applications, micro pumps to develop positive fluid pressures, micro flaps to modulate air streams on airfoils and so on. MEMS sensors are widely used in biotechnology products such as enzyme linked immunosorbent assay (ELISA), capillary electro-phoresis, electro-poration, scanning, tunnelling microscope biochips for reduction of hazardous chemical and biological agents, and micro systems for high-throughput drugs screening and selection. In medicine a number of devices are MEMS based. Other applications include communication and inertial sensing.

In defence applications there is a paradigm shift in sensors used for observing the enemy's war resources, force deployment and their tactical manoeuvres in the battle space. The future sensors would harness the MEMS technology and nanotechnology to sense threats. They would provide modularity and flexibility in the intelligence, surveillance and reconnaissance (ISR) operations. Satellites, UUVs, UAVs would become the main sensor platforms and multi-sensors would be deployed for getting composite picture of the battlefield through information fusion. Future wars would employ extremely high-resolution sensors so that the information could be fed to the weapon guidance and control systems to enable the weapons to hit only the right target and eliminate the threat. Highly sensitive passive MEMS with long detection ranges, improved probability of detection, minimum false alarm rate, and better resolution would be used in the future applications. These sensors would also have anti-jamming features, anti-stealth capabilities and multi-static and network-centric configurations. Advanced digital signal processing techniques would be incorporated to provide real time information about the enemy. The typical applications of MEMS for defence are shown in Figure 3.34.

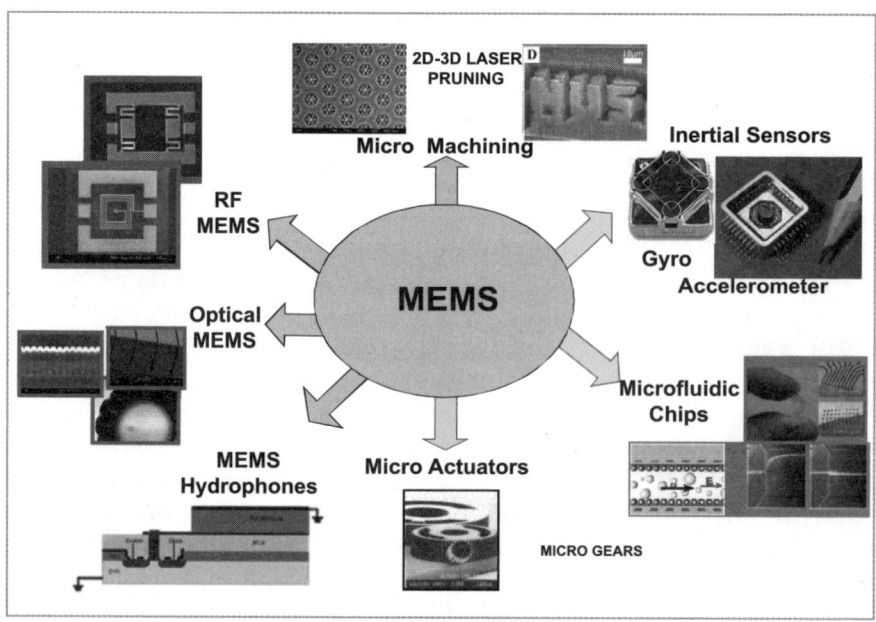

Fig 3.34: MEMS Applications

Sensors for Detecting IEDs

A wireless sensor network is a collection of sensor nodes that are organized into a cooperative network; they are "ad-hoc systems" containing sensors connected by wireless links (Akyildix et al, 2002). Wireless sensor networks have numerous applications, ranging from habitat monitoring, environmental control, to the military realms of intelligence, surveillance and reconnaissance (ISR). Sensor networks have the advantage over other surveillance technology of a widely distributed presence, minimal intrusiveness, and minimal need for human interaction (Haenggi, 2005). Detection of IEDs could benefit from them because sensor networks are increasingly used to cooperatively detect and identify targets of interest. Sensors nodes picking up suspicious activities could forward the data via repeaters to relay stations and notify backend operators and analysts, who could activate services such as the rescue teams, fire brigades, and law enforcement agencies. Wireless sensor networks appear promising for detecting IED emplacement because they can provide uniform coverage of a wide area. Surveillance by camera provides only a limited number of perspectives and can suffer from occlusion problems. Furthermore, sensors of a variety of different modalities can provide robust and more accurate detection of IEDs (Tran et al, 2007) than can perform imaging alone. Diffuse passive infrared and magnetic sensors appear good at finding suspicious behaviour of both people and vehicles (Caruso and Lucky, 2007), they can be cheaper than cameras, and the necessary data processing can be simpler than that for images, magnetic sensors are particularly good at vehicle detection (Rouse et al, 1995) and have been used for detection of unexploded ordnance (Wiegert & Oeschger, 2006). Diffuse passive infrared detectors are a mature technology with many applications, and just a few infrared sensors can accomplish complex monitoring tasks (Kaushik, Lovell & Celler, 2007). There are additional problems with processing to track and detect suspicious behaviour which we have addressed in previous work (Rowe, 2005).

Command and Control of IED Detection

Since IED detection requires a wide range of techniques, coordination of efforts is essential by a carefully designed command-and-control structure. Priorities and deployment parameters need to be assigned to the techniques. For instance, one must decide how valuable it is to use chemical sensors to detect explosives versus putting detectors on booms in front of vehicles versus doing video surveillance. This analysis then determines the allocation and deployment of personnel and equipment.

Sensor network can be considered as an extension of a military command-and control hierarchy where devices rather than soldiers are patrolling. With a varied set of sensors, instructions analogous to orders are given as to what information is to be collected. With software control, instructions can be changed quickly as the situation in the sensor field develops. IED detection methods particularly need to be reviewed periodically as enemies adapt. Adaptation is an important phenomenon today in Iraq, for instance IED emplacers quickly shifted in the course of a year from radio-wave triggering to command-wire triggering (Atkinson, 2007). This means the clues to look for in IED emplacement change too, since we can now look for evidence wires being laid to the device. It is thus helpful for IED detection to focus on general-purpose knowledge of the observed behaviour such as tracks and accelerations rather than trying to identify specific signatures of an IED emplacer. That has been our strategy here.

3.4.5 Photonics

Many new areas have come up during the last few decades e.g. Electro-optics, opto-electronics, light wave technology, etc., which in short are called Photonics. The range of application of photonics starts from energy generation, energy detector to communication and information processing. One of the most important applications of photonics is in the area of fibre-optics communication. Photonics has found important military applications that include improved data communication and telecommunication for command and control systems, C^4I^2SR systems, displays and electronic warfare systems. It also has application in areas like sensors, fibre-optic gyros, fibre-optic guided missiles and fly-by-wire flight control systems. Photonics is therefore one of the most researched subjects in the world today. DRDO has initiated a major Photonics programme to develop emerging technologies and indigenise critical technologies (Figure 3.35). The key technologies successfully developed are Adaptive Optics technology which is being used in directed energy weapons and pointing and tracking of space satellites, Integrated optic components and devices for high speed signal encoding and for sensor signal processing, Photonic correlators for Automatic Target Recognition by Armed Forces, Diode pumped solid state laser, Optical phase conjugation technique for laser beam delivery system etc.

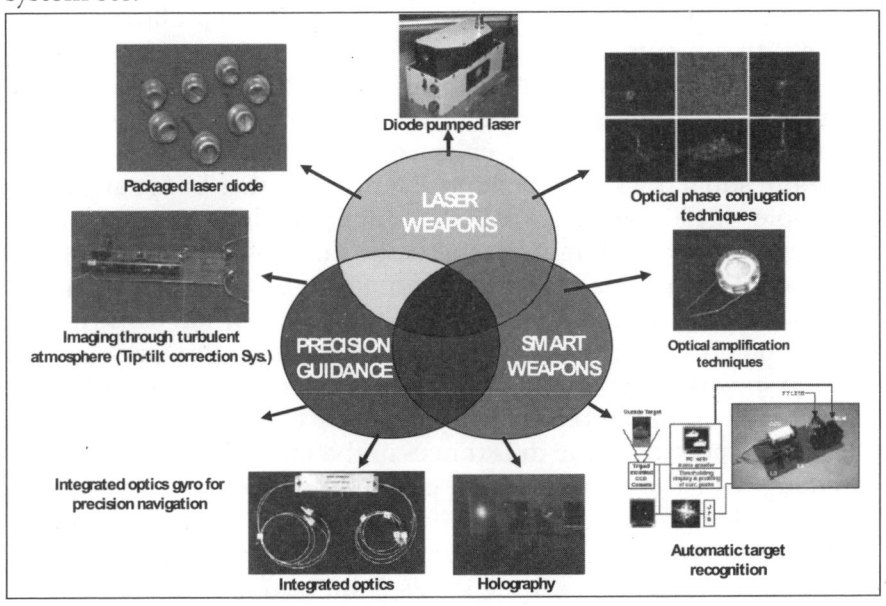

Fig. 3.35: Photonics - Development Thrust in India

Some of the important developments in Photonics and related technologies with potential applications for guided missile systems include: Uncooled Imaging technology, Focal plane array technology, Conformal optics, Adaptive optics, Fiber optics technology, Micro-Opto-Electro-Mechanical Systems (MOEMS), Optical Correlators, Multi-Spectral fusion (fusion of 3-5 μ and 8-12μ images), Multi-sensor Fusion (fusion of 3-5μ and 8-12μ images of IR and MM wave images), Real-time Image Processing etc.

Photonics have also got a major role to play in other fields. Biophotonics is the science of generating, transmitting, and detecting photons that interact with tissue or biomolecules and has come a long way in recent years. Biophotonics has the potential to provide enabling technologies for significant advances in medical diagnostics, therapeutics and biotechnology by exploiting light absorption and re-emission, as well as elastic and inelastic photon scattering events in tissues or samples. Photons can travel through tissues, probing the tissue through various light-tissue interactions and eventually carrying the information back to the tissue surface for subsequent optoelectronic detection. Photonics techniques are increasingly being applied in medicine. For example, the interaction of light with human and biological cells is used to non-invasively detect specific tissues such as tumours. Light can also be used to change the properties of certain tissues and cells allowing non-invasive treatments. DNA can be detected and sequenced using biophotonic methods. Photonics techniques are used extensively in medicine to improve visualization techniques. Laser-induced fluorescence is used for early detection of cancer in the gastrointestinal track and the lung. These and other body cavities are accessible via flexible endoscopes enabling the delivery of excitation light, commonly blue light, to the tissue and collecting the emitted fluorescent light, usually in the green and red wavelength band, by fibre bundles for imaging outside of the body. Normal tissue can be differentiated from pre-cancerous and cancerous sites based on the emitted spectrum or colour of the light, where the latter sites emit less light overall and preferentially at longer wavelength, e.g., in the red part of the spectrum. The differences in the emission spectra are due to changes in the anatomy and presence of different fluorescing biomolecules. Thus photonics have a major role in medical diagnosis.

Transition in lighting from incumbent technology to low energy consumption, digital technology, is built around LEDs, OLEDs, sensors and microprocessor intelligence. The introduction of advanced photovoltaics, low-energy light sources and intelligent lighting controls will lead to substantial reductions in lighting energy requirements. Photonic sensing and imaging will contribute to a greener environment by advanced pollution detection, and enable higher levels of security and safety through the use of sophisticated surveillance technology and detection of unauthorised goods.

Bio-sensor

Biosensor is a small device employed for biological recognition properties for a selective bio-analysis. It uses biological material to conduct and monitor the presence of specific chemicals in an area. This eliminates the need for sample preparation and analyzes the sensing elements that may be whole-cells, anti-bodies, enzymes or DNA.

There are two elements: the sensing element is responsible for the selective detection of the analyte, the transducer converts a chemical event into an appropriate signal that can be used to determine the analyte concentration in a given test sample containing the sensing elements. Biosensors are of interests due to tremendous promise for obtaining sequence specific information in a faster, simpler and cheaper manner compared to the traditional analysis.

Nanotechnology is playing an increasingly important role in the development of biosensors. The sensitivity and the performance of biosensor are being improved by using nanomaterials for their construction. The use of these nanomaterials has allowed the introduction of many new signal transfusing technologies in biosensors. Because of their submicron dimensions, nanosensors, nanoprobes and other nanosystems have allowed simple and rapid analyses.

Nano Structure in Biosensors

Acoustic wave particle, magnetic and electro-magnetic biosensors will have numerous applications using nanomaterials.

Carbon nanotube: Carbon nanotube is one which is a significant application in bio- sensors. Carbon nanotube (CNT) exhibits a unique combination of excellent mechanical, electrical and electromechanical properties. This has stimulated increasing interest in the application of CNT components in bio sensors and CNT based bio sensors for futuristic applications in biology and medicine. The application includes non-invasive sensors for glucose and cholesterol level measurement.

Porus silicon: Nanostructure material that has been studied extensively for nano sensing applications is nanocrystalline silicon, often referred to as porous silicon. Since the discovery of its strong visible luminescence at room temperature, porous silicon has attracted considerable interest in its possible use in the construction of biosensors. Its ability to emit light is due to its tiny pores that range from less than 2 nm to micrometer dimensions.

Electrochemical Sensors

In this configuration, sensing molecules are either coated onto or covalently bonded to a probe surface. A membrane holds the sensing molecules in place, excluding interfering species from the analyte solution. The sensing molecules react specifically with compounds to be detected, sparking an electrical signal proportional to the concentration of the analyte. The bio-molecules may also respond to an entire class of compounds and the detection method for electrochemical biosensors, involves measurement of current, voltage, conductance, capacitance and impedance.

Optical Ssensors

In optical biosensors, the optical fibers allow detection of analytes on the basis of absorption, fluorescence or light scattering. Since they are non-electrical, optical biosensors have the advantage of lending themselves to in vivo applications and allowing multiple analytes to be detected by using different monitoring wavelengths. The versatility of fiber optics probes is due to their capacity to transmit signals that report on changes in wave-length, wave propagation, time, intensity, distribution of the spectrum, or polarity of the light. In general, acquisition of the signal from these devices is accomplished through flexible cables which can transmit light to the biological component.

Optrodes use fiber optics for performing optical measurement away from the measuring locations (e.g., intra-arterial determination using FIA systems). A powerful and sensitive analytical methodology has been constructed based on the luciferin/ luciferase bioluminescene reaction.

Piezoelectric Sensors

In this mode, sensing molecules are attached to a piezoelectric surface – a mass to frequency transducer in which interactions between the analyte and the sensing molecules set up mechanical vibrations that can be translated into an electrical signal proportional to the amount of the analyte (Figure 3.36). An example of such a sensor is quartz crystal micro or nano balance.

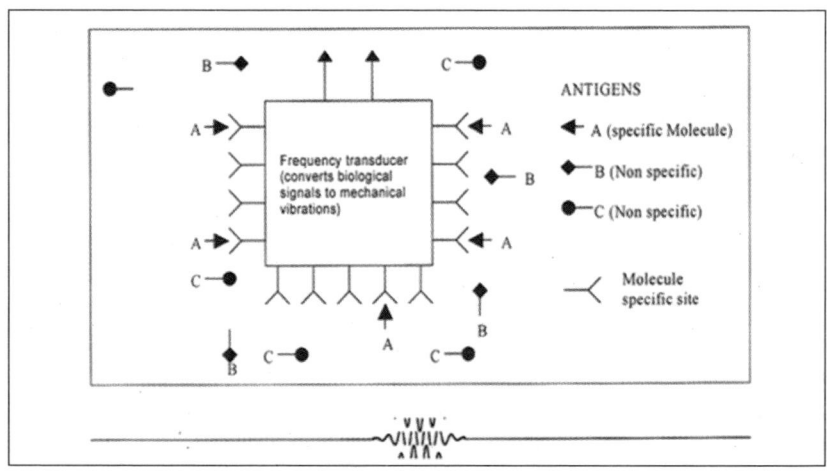

Fig. 3.36: Piezoelectric Sensor

Field Effect Transistor (FET)

This method makes use of an ion-sensitive field effect transistor (ISFET) built on standard technology that produces source, drain and gate regions. The gate uses an ion sensitive membrane that renders ISFET capable of biochemical recognition in the presence of the analyte with an increase in local ion concentration.

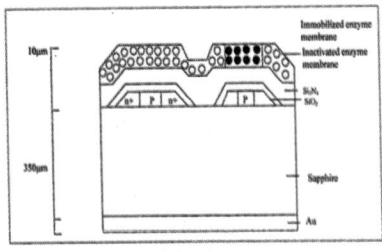

Fig. 3.37: Ion-sensitive Field Effect Transistor

Microelectrodes are created on a silicon nitrate surface using vapour deposition method and partially insulated by titanium oxide (Figure 3.37). The hardware component consists of an electrode system that could either be a conventional platinum or silver-silver chloride microelectrode and a field effect transistor with an ion sensitive gate or gas sensing electrode.

Detection of Biological Agents

The present day technology for detecting bio-warfare agents of different bacteria, toxin viruses which have led to antibodies is the recognition molecule which binds with a service feature of the agents or has used nucleic acid and determined a nucleic acid sequence known to be found in the agents being tested.

3.4.6 Imaging Sensors

Uncooled thermal imaging systems are very important for missile systems as these systems offer significant operational advantages and saving of weight, space as well as cost associated with the cooling system. Towards this, the resistive microbolometer technology, ferro-electric bolometer technology, uncooled thermo electric linear arrays are becoming more important. The research effort in uncooled infrared technology is directed towards several other military applications including reconnaissance, surveillance and weapon sighting capabilities as well as for precision munitions and dispenser system applications and anti-armour sub-munition programmes. Focal Plane Arrays technology is crucial for missile seekers and other imaging sensors. The focus is on developing larger size arrays, higher resolution and higher sensitivity focal planes. The current research is also focused on reducing the pixel size and increasing pixel sensitivity using advanced materials and micro electromechanical device structures. For example, target detection and lock on range of a third generation anti-tank guided missiles can be significantly increased by enhancing the performance of the FPA. Adaptive optic system is a growing area of interest for the guided missiles. Adaptive optic systems typically consist of a wave front phase sensor, focusing optics, a spatial light modulator (SLM) for correcting phase errors, imaging sensors and control and processing electronics. These systems improve the image quality by reducing the phase aberrations introduced when the wave front travels through turbulent atmosphere or aberrations introduced by the optical system itself. Advanced technologies are now becoming available to make these systems lightweight, low power and compact. The technologies that are making this possible include highly integrated low power electronics, and new processing architectures for error sensing and control, flexible high density packaging and Micro-Opto-Electro-Mechanical Systems.

3.4.7 Lasers

LASER stands for Light Amplification by Stimulated Emission of Radiation. This refers to a stream of light particles called photons getting amplified on a laser active medium (Figure 3.38). The wavelength of the generated photons can be modified depending upon the condition of laser active medium – Gas, Liquid or Crystal.

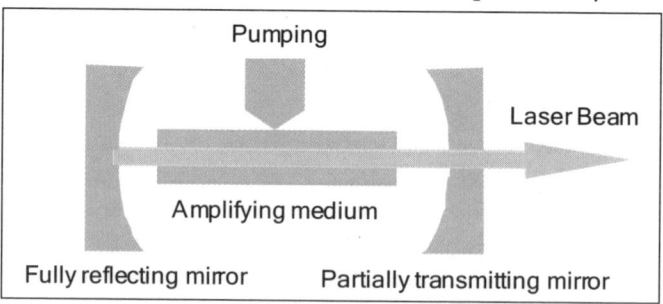

Fig. 3.38: Simplified Schematic of Typical Laser

The laser beam is highly directional with high intensity and very high frequency or wavelength. It comes in sizes ranging from one tenth of a cross-section of a human hair to the size of a very large structure, emitting powers ranging from 10^9 to 10^{20} watts and web lengths ranging from microwave to soft-x-ray spectral regions ranging from frequencies of 10^{11} to 10^{17} hertz. The pulse energy of a Laser is 10^4

Joules and pulse duration is as short as $6x\ 10^{-15}$ sec.

Based on the wavelength, the laser beams have been categorised as gas laser, excimer laser, solid state and diode laser. The nature of the emission is a continuous waveform or pulsed mode. These lasers have wavelength in visible, infrared, ultraviolet, microwave and x–ray regions. The lasers are used for material processing, medical technology, measurement, communications, inertial sensors and other industrial & security applications.

The Excimer Laser is for semi conductor photo lithography and for eye surgery. Solid state laser is used for material processing, medical technology measurement and communication. It can also be used for tele-communication. Diode laser is again applicable for many uses, especially, Laser spectroscopy.

Applications

Lasers have become an integral part of our lives. Lasers have a variety of applications in medical, welding and cutting, surveying, communication, heat treatment, barcode scanning, laser cooling etc. Lasers are used for photocoagulation of the retina to halt retinal haemorrhaging and for the tracking of retinal tears. Higher power lasers are used after cataract surgery if the supportive membrane surrounding the implanted lens becomes milky. Photo disruption of the membrane can often cause it to draw back like a shade, almost instantly restoring vision. A focused laser can act as an extremely sharp scalpel for delicate surgery. They are also used in CD/DVD drives, laser printers, surgical and clinical equipment, various flow measurement equipment, remote sensing and in holography, communication, alignment and levelling instruments, mapping human genome, measuring the distance between the earth and moon, etc. The field of application is expected to grow in the future as the scientific community moves towards the realization of quantum computers, quantum communication, 3-D data storage devices, terahertz networks etc.

Blue Laser: Compact and efficient solid state and diode laser with a motion wavelength in the blue and ultra violet spectral range is called blue laser.

The importance of blue laser is in the area of micro technology and micro material processing. From a small size structure like computer processor chips required for precision fabrication, they are also used in the field of data storage and holographics. This allows to double the storage capacity of a CD.

Green Laser: Living cells can be genetically engineered to produce green fluorescent protein (GFP). The GFP is used as the laser's 'gain medium', where light amplification takes place. The cells are then placed between two tiny mirrors, just 20 millionths of a meter across, which act as the 'laser cavity' in which light can bounce many times through the cell. Upon bathing the cell with blue light, it can be seen to emit directed and intense laser light.

Classification:

Category wavelength	Application
Gas laser (ultraviolet, visible, infrared)	Material Processing Medical Technology
Excimer laser (ultraviolet)	Semiconductor photo lithography
Solid-state laser (visible, infrared)	Material Processing Medical Technology Measurement Technology Optical Telecommunications
Diode laser (visible, infrared)	Material Processing Medical Technology Measurement Technology Optical Telecommunications Everyday Appications
Dye laser (ultraviolet to near infrared)	Laser Spectroscopy

Ref.: Laser Fundamentals by William T. Silfvast

High Power Lasers

Higher power lasers are used generally as weapons called directed energy systems. This includes guns, missiles and bombs which destroy the target with kinetic energy. Two of the most fundamental problems seen with projectile weapons are getting the projectile to successfully travel a useful distance and hit the target producing useful damage effects, and problems shared by direct energy weapon. Weapons technologies include high energy laser weapon (HEL), high power microwave (HPM), particle beam and laser induced plasma. HELs have the greatest potential to produce significant effect.

Carbon dioxide (CO_2) lasers with powers up to several kilowatts are used for computer controlled welding on auto assembly lines. CO_2 laser can weld stainless steel with copper. Great difference in thermal conductivities between these two metals makes conventional welding near impossible. Helium-Neon and semiconductor lasers have become standard parts of the field surveyor's equipment. A fast laser pulse is sent to a corner reflector at the point to be measured and the time of reflection is measured to get the distance. Pulsed Ruby laser is used for surveying long distance. A technique that has recent success is laser cooling. This involves atom trapping, a method where a number of atoms are confined in a specially shaped arrangement of electric and magnetic fields. Particular wavelengths of laser light at the ions or atoms slow them down, thus cooling them.

Military uses of high power lasers include applications such as target designation and ranging, defensive counter-measures, communications and directed energy weapons. Defensive counter-measure applications can range from compact, low power infrared counter-measures to high power, airborne laser systems. IR countermeasure systems use lasers to confuse the seeker heads on heat-seeking anti-aircraft missiles. Chemical laser powered by an energetic chemical reaction is used as the main weapon beam to find, track and destroy incoming long range ballistic missiles. Laser guided bullets can steer themselves on their own and home on to targets. These bullets are 10 cm long with fins and a little optical sensor in the nose that can lock onto a laser target designator. The electronics and

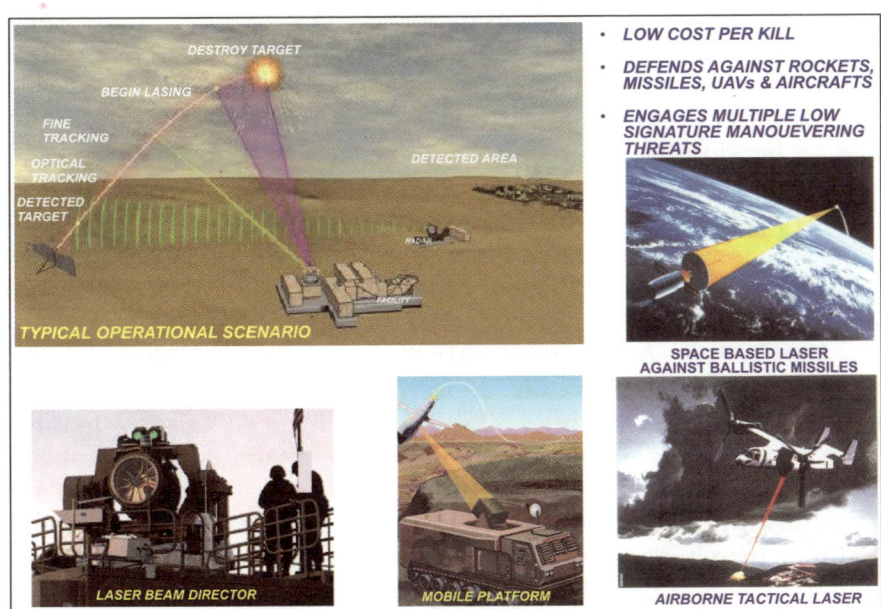

Fig. 3.39: Directed Energy Weapons

actuators inside the bullet enable the fin movement and steer the bullets in-flight. The Mobile Tactical High-Energy Laser and the airborne laser are in use in some countries as defensive laser system. This field-deployable weapon system is able to track incoming artillery projectiles by radar and destroy them with a powerful deuterium fluoride laser. Application of High Power Lasers and High Power Microwave technologies is shown in Figure 3.39. This is the new class of "Speed of Light" weapons that are going to be of high strategic importance in the coming century.

Chemical Oxygen Iodine Laser (COIL)

COIL is an infrared chemical laser with output wavelength of 1.315 μm and transition of atomic iodine. The laser is fed with gaseous chlorine, molecular iodine, and an aqueous mixture of hydrogen peroxide and potassium hydroxide. The aqueous peroxide solution undergoes chemical reaction with chlorine producing heat, potassium chloride, oxygen in excited state, and singlet delta oxygen. Spontaneous transition of excited oxygen to the triplet sigma ground state is forbidden giving the excited oxygen a spontaneous lifetime of about 45 minutes. This allows the singlet delta oxygen to transfer its energy to the iodine molecules injected to the gas stream; they are nearly resonant with the singlet oxygen so the energy transfer during the collision of the particles is rapid. The excited iodine then undergoes stimulated emission and lasers at 1.315 μm in the optical resonator region of the laser.

The laser operates at relatively low gas pressures but the gas flow has to be close to the speed of sound at the reaction time; even supersonic flow designs are described. The low pressure and fast flow make removal of heat from the lasing medium easy in comparison to high-power solid-state lasers. The reaction products are potassium salt, water, and oxygen. Traces of chlorine and iodine are removed from the exhaust gases by a halogen scrubber.

COIL has numerous other commercial applications. Experimental COIL cutting results; estimates from a theoretical model determined that a 10–30-kW fibre delivered COIL would meet the needs of the decommissioning and decontamination of nuclear facilities. Other potential industrial applications for COIL are shipbuilding, automotive manufacturing, heavy machinery manufacturing, and tasks requiring underwater cutting or welding, and there may be useful applications in the oil and gas industry.

Electro Magnetic Pulse

High power Electromagnetic pulse is also a directed energy weapon which can be deployed from unmanned aerial vehicles or missiles towards any specific targets. This is a non lethal weapon and will not damage any human being or structure or  equipment. The directed energy waves will neutralize all the communication, computers and power systems without the knowledge of the operators. This has been experimented successfully by the Boeing under the programme CHAMP (Counter-electronics High powered microwave Advanced Missile Project). This project is a cruise missile that replaces an explosive warhead to electro magnetic pulse called e-bombs which can transform modern warfare to totally non-lethal but very effective in neutralizing enemy's assets.

3.5 MATERIALS AND PROCESSING

The demand for materials is ever increasing as new applications are emerging with particular specifications and quality. It is also true that the raw material resource is limited and is getting depleted day by day. Therefore, more and improved technologies will be needed in the future so that specific consumption is minimized for every product and at the same time the materials are of different specifications. Inventions continue in many research labs to find those new materials. Figure 3.40 shows different materials and their applications. These materials include metal alloys, composites, ceramics, polymers, semi-conductors, bio-materials, explosives, nano-materials. Each of these materials has many applications in civilian, military and space programmes. It is a testimony that finer aspects and further research on these materials could offer much more. Smart materials are becoming prominent due to their adaptive shape and behaviour according to the specific environmental conditions, thus reacting intelligently. These adaptive materials with embedded sensors will mark a revolution in material technology. Simple building blocks can arrange themselves to create units of higher complexity with different properties as self-organising systems.

Light weight structures are of high demand for aerospace programmes. Composite materials are used for rocket motor casings and also as ablative liners with various processing techniques such as

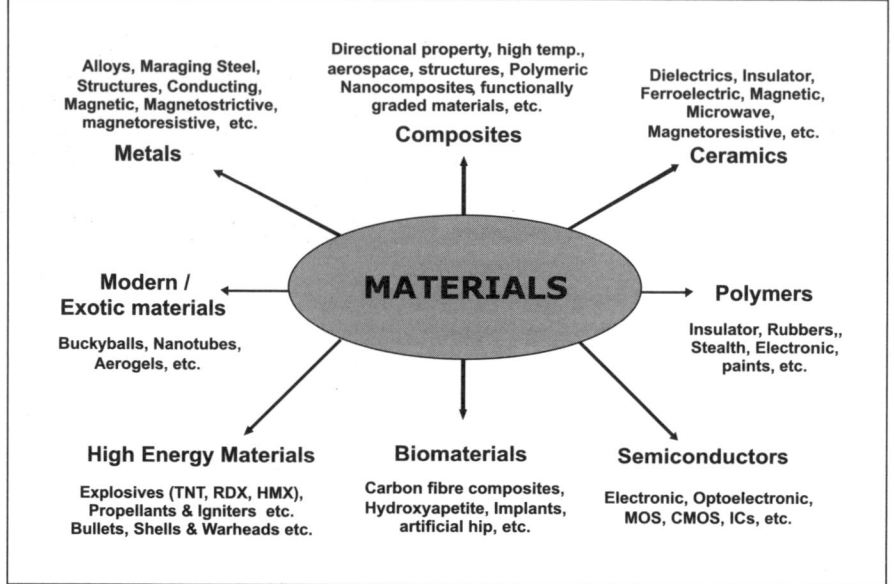

Fig. 3.40: Materials and Applications

filament winding, tape winding, autoclaving, etc. Stealth properties of materials are also important particularly for defence applications in fighter aircraft, missiles, etc. The recent technologies are related to adaptive camouflaging and meta-metals which provide adequate stealth against deduction with respect to environment. Recent developments on active and smart materials have a major role in the future in developing adaptive aerospace systems like morphing wings of an aeroplane which is able to take off without a runway like a bird. Details are brought out in the succeeding paragraphs.

3.5.1 Stealth Technologies and Materials

For defence applications it is needed to deceive the detection from enemy sensors and instruments. The detection methods are generally through visibility of the naked eye, infrared, TV camera, radars, etc. So the system should be adapted in such a way that none of the detection methods can work. The technology used for non-detection is stealth. This can be achieved by choice of material for the surface, special coatings, shape to give low radar cross sectional area.

Stealth technologies and materials are essential for the survivability of weapon systems such as missiles, fighter aircraft, armoured vehicles, ships, submarines etc., to protect them from enemy attack. Suitability of camouflaging materials as well as design patterns is decided based on spectral, spatial and temporal variation of the background signature of the terrain. Special paints have been developed to match the spectral characteristics of the terrain so that it becomes difficult to visually detect armoured vehicles or any other object which has been painted.

Stealth materials and technologies to give protection from electromagnetic (EM) waves use EM wave absorbing materials such as carbon fibres and ferrites as reinforcement in advanced composite structures. Ferrites are quite popular and they are capable of absorbing EM signals to the extent of power loss of – 40dB, therefore, nothing is reflected back to radar to detect the object. However, the high volume fraction of ferrite in advance composites is a challenge because higher volume fractions of ferrite cause brittleness in the composite structure. Nanotechnologies have also come to the fore of stealth technologies. Nanoferrites hold a lot of promise as future stealth materials. Development of more advanced radar absorbing materials and techniques to suppress infrared signature is essential to enhance the scope and characteristics of stealth materials and technologies. Suppressing the infrared signature is important for camouflaging of armoured vehicles such as main battle tanks, infantry combat vehicles, etc. New stealth technology such as electronic ink has been reported for Battle Tanks to make them completely invisible by merging its colour with the background. The electronic ink works instantly and provides complete camouflage from visual ranges.

3.5.2 Material for Adaptive Camouflage

For certain applications, materials need to be adaptive to the environment. These materials are photochromic, magnetochromic and luminescent nanotubes and nano-fibres. These materials come under the category of smart and intelligent materials and manifest different properties when subjected to light, magnetic fields. These materials are smart enough to sense the surrounding area and its pattern and intelligent enough to bring about a suitable change in their appearance in order to merge with the surrounding pattern and environment.

Nano-inorganic materials are excellent material for radar absorbing coatings and structures and also for adaptive clothing. Use of these will further augment survivability.

3.5.3 Active and Smart Materials

The smart materials respond to a predictable action and have excellent repeatability of the response on a stimulus. Their response is well-defined and controlled, whereas, normal materials have limited response. Piezoelectric materials made a remarkable impact on human life ranging from simple industrial equipment to healthcare systems. The examples of effects of smart materials are piezoelectric, shape memory materials, photochromic, electroluminescent, thermochromic, etc. (Figure 3.41). Piezoelectric effect describes the relation between a mechanical stress and an electrical voltage in solids. An applied mechanical stress will generate a voltage and an applied voltage will change the shape of the solid by a small amount (up to a 4 per cent change in volume) and it is reversible. The piezoelectric effect occurs only in non conductive materials. Crystals and ceramics are the examples of piezoelectric materials. Photochromic materials change colour reversibly with changes in light intensity. Changes from one colour to another colour are possible mixing photochromic colours with base colours. They are used in paints, inks, and mixed to moulding or casting materials for different applications. The shape memory alloys (SMAs) are another category of active and smart materials. SMAs such as Ni–Ti, Cu–Al–Zn, etc., may undergo large strains and recover their initial configuration spontaneously at the end of the deformation process or by heating without any residual deformation. Currently, SMAs are mainly applied in medical sciences, electrical, aerospace and mechanical engineering and can also open new applications in civil engineering, specifically, seismic protection of buildings.

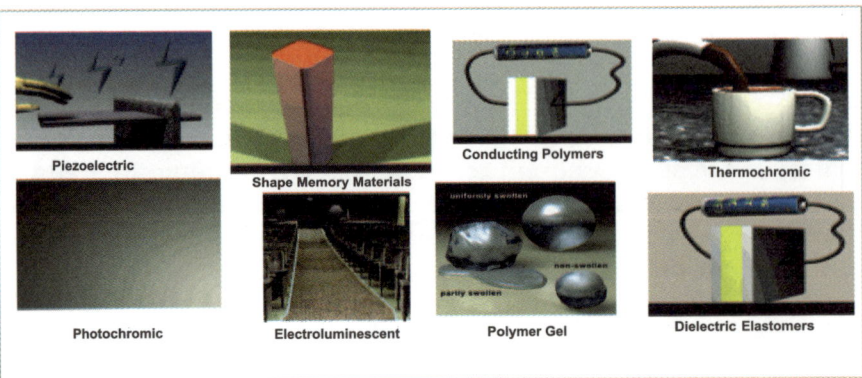

Fig. 3.41: Smart Materials

Similarly, smart concrete has a lot of potential in civil construction. A mere addition of 0.5 per cent specially treated carbon fibres enables the increase of electrical conductivity of concrete. A mechanical load, on this concrete reduces the effectiveness of the contact between each fibre and the surrounding matrix thus slightly reducing its conductivity. On removing the load the concrete regains its original conductivity. Because of this peculiar property the product is called 'Smart Concrete'. The concrete could serve both as a structural material as well as a sensor. The smart concrete could function as a traffic-sensing recorder when used in road pavements. It has got higher potential and could be exploited to make concrete reflective to radio waves and thus suitable for use in electromagnetic shielding. The smart concrete can be used to lay smart highways to guide self steering cars which at present follow tracks of buried magnets. The strain sensitive concrete might even be used to detect earthquakes.

3.5.4 Invisibility through Meta-metals

Meta-metals is an engineered material made by embedding circuit elements in array-like fashion into a dielectric substrate for high level stealth. This leads to a negative refractive index across the materials (Figure 3.42). These materials provide protection against microwave radiation and thermal imagers and create a shield against visibility thus providing total invisibility. The surface of meta material provides

Meta Material is an engineered material, made by embedding circuit elements in an array-like fashion into a dielectric substrate for high level of stealth

• *Electric permittivity* is negative
• *Magnetic permeability* is negative

This leads to negative refractive index of the material.

Cloak technology provides protections:
- Against microwave radiation
- From Radar observing
- From Thermal imager
- Creating invisibility to human eye

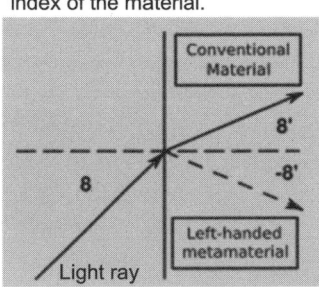

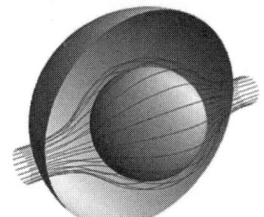

One way of cloaking is to make light flow around it like a stream of water using negative refractive index of the material

Fig. 3.42: Invisibility through Meta-metals

flow of light as if water is flowing over the surface and, therefore, obstruction due to the presence of opaque material is not observed.

3.5.5 Advanced Composites

Composites are quasi-homogeneous materials consisting of two or more materials which are insoluble in each other, to create properties not achievable individually. This process is called reinforcement to improve the mechanical properties of a component i.e. higher strength with less weight. The material groups that come under the composites are polymer compound materials called Fibre Reinforced Plastics (FRP) – Thermosets, Thermoplastics and Elastomers, Metal Matrix Compounds (MMC), and Ceramics Matrix Compounds (CMC) – Silicon Carbide, Silicon Nitride, etc. High performance fibre composites are generally carbon fibre based or glass fibre based reinforced polymers. These fibres are impregnated with resins to form textile semi-finished materials like fabrics or fibres which are used either in the process of winding or moulding to take the required shape and then cured in ovens. Application of the composite material in light combat aircraft and missiles is discussed ahead.

Composites in Light Combat Aircraft (LCA)

Materials distribution in LCA is carbon fibre reinforced composites (CFC) – 44 per cent, aluminium – 43 per cent, titanium & steel – 5 per cent and other materials – 8 per cent (Figure 3.43). The design of CFC wing has adopter and bolt technology where the top and bottom skins are fastened to the substructure consisting of CFC spars and metallic boundary members. The CFC spars have been made using moulds with pressure pad for proper consolidation at corner regions. This has minimised and simplified the use of shimming on assembly. The design methodology follows an optimisation exercise to get optimum thickness distribution in various regions of the skins, using ELFINI software. It then goes through a skin engineering cycle involving determination of stacking of different plies of prepreg with fibres in different directions based on stacking sequence design. The detailed ply-by-ply drawings have been made using the AUTOLAY software.

Some of the noteworthy features of the design and fabrication of the CFC wing are: boundary members, the nose box, slats and the structure in chord-wise direction along the two pylon stations which are made of aluminium alloy. All fasteners used for assembling sub-structure and one in five fasteners assembling skins to sub-structure

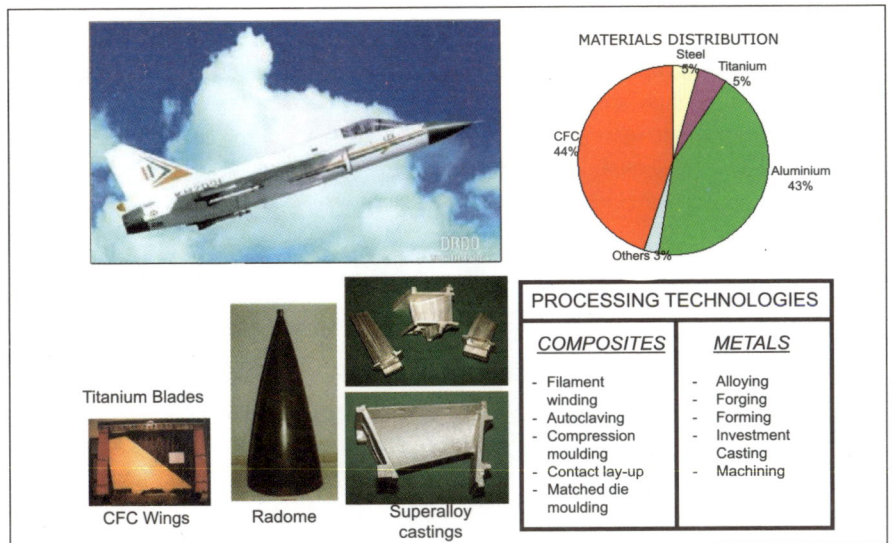

Fig. 3.43: Materials Distribution in LCA - Tejas

are dipped in 'Co-Bond-2160' (a conductive sealant) to avoid sparking at fastener heads during lightning strikes. The internal system conduits are assembled using heat shrinkable metallic sleeves (lightning insulators). All fastener heads on the top and bottom skins are covered by aluminium foil strips with a polyester film in between to minimise accumulation of charge on skin.

Composites in Missiles

Missile projects in India brought out many new technologies using composite materials. Carbon-carbon composites, airframe sections, rocket motor casing, rocket motor liners, etc., were made using a different variety of advanced composites. Materials transparent to electromagnetic radiation with structural and thermal stability are important to aerospace industries. Such materials are ceramics and advanced composites. The primary requirement of such materials is to have low values of dielectric constant and tan-δ. Toughened ceramics made by sol gel processing technique and Kevlar or glass reinforced plastics are the materials which are currently in use. Polymeric nanocomposites are materials which have enormous potential for the future. Nanoparticles/Nanofibres reinforced polymeric components are promising materials for the future. Presently, there is a need to work on cost effective technologies to exploit its full potential through commercialization.

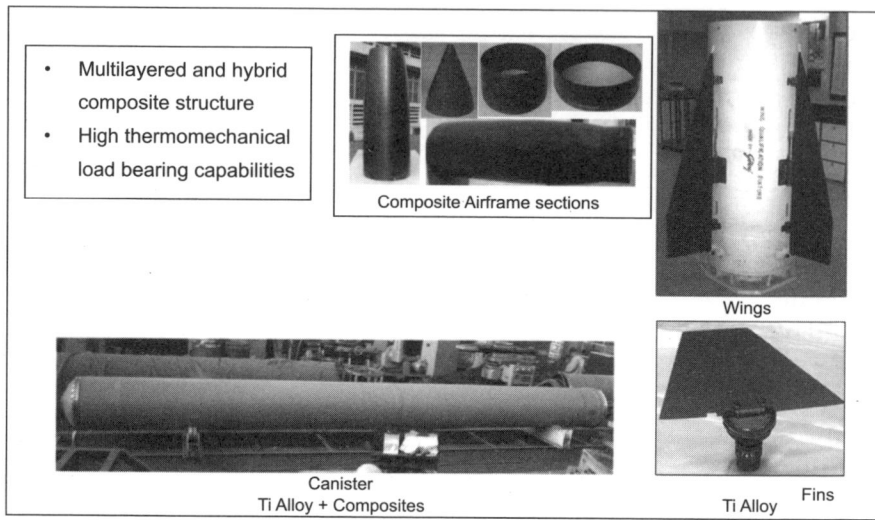

Fig. 3.44: Airframes for BRAHMOS

Multifunctional Composite Airframes

BRAHMOS missile uses the synergy of technological competence and a consortium of the industries of both partner countries. Advanced high temperature composites and hybrid composites with multi-functionality have been used in BRAHMOS missiles. The new processing technologies and multi reinforcement with multi-matrix system were evolved for the first time for the development and manufacture of this missile (Figure 3.44). The composite smart structures in BRAHMOS are multifunctional and cater to many assigned tasks at a time.

The industry consortium established with active participation from both public and private industries from India and Russia has also played a pivotal role in the success of this development. It has definitely set an ideal example for collaborative ventures amongst partners from other countries with India.

3.6 HIGH ENERGETICS

3.6.1 Explosives and Armaments

The first explosive was known as gunpowder or black powder which was a mixture of charcoal, sulphur and potassium nitrate. Initially, it was used to create fireworks and later on it was used as an explosive to shoot stones and spear-like projectiles from tubes, and metal balls from cannons and guns. The exact potential of explosives was first acknowledged with the invention of nitroglycerine in 1846 and later with the invention of dynamite in 1860. Later on, as technologies advanced, newer varieties of explosives were invented and TNT, RDX, HMX, etc. were invented in the early twentieth century which also saw the full exploitation of these explosives in many wars including world wars.

Later by the middle of the twentieth century, massive ordnance was developed and the mother of all bombs was made with 11 tonnes of explosive and at the beginning of the twenty first century, the father of all bombs with 44 tonnes of explosive was also developed. Presently, very high energy explosives such as CL20 (Hexanitrohexaazaisowurtzitane), FOX-7 (1,1-Diamino-2,2-dinitroethene), TNAZ (1,1,3 Trinitroazitidine), TATB (triamino-trinitrobenzene), are in demand for highly efficient weapons. In the area of explosives, phenomenal growth has been registered all over the world. Most of the explosives find applications in armaments and ammunitions. India has mastered the use of armament systems and explosives. Multi Barrel Rocket (MBR) System (Pinaka) is an area weapon with a range of 10 to 38 km. It has got a quick reaction time of 3 sec. It carries a 100 kg warhead and fires a salvo of 12 rockets in 44 sec. High performance explosive used in MBR is capable of neutralising a target area of 1000 x 800m.

Kinetic Energy (KE) explosive used in Fin Stabilised Armour Piercing Discarding Sabot (FSAPDS) with tungsten alloy penetrator and semi-combustible case is capable of defeating single, double and triple heavy standard targets at 2,000m range. Pre-frag Bomb consists of a Dentex filled pre-fragmented Fibre Reinforced Plastic (FRP) module with three layers of steel balls capable of penetrating 6mm MS Target up to 40m from the point of burst. A versatile tube launched laser guided missile has a tandem HEAT (High energy antitank) warhead and can defeat advanced main battle tanks (MBTs) and also engage both ground and aerial targets at ranges up to 2 km. Fuel air explosives are new generation and lightweight explosives. They use atmospheric oxygen for exothermic chemical reaction to cause explosion, therefore, the efficiency of fuel air explosives is very high.

Future explosive systems are integrated systems which involve various state-of-the-art technologies. With new technology development all over the world, any system in isolation cannot perform beyond a limit. It is essential to synergise the output of a system to get best efficiency and accuracy. The same is true with future explosive systems.

Modern munitions are effective in their military capabilities and safe despite usage of energetic materials like high explosives and gun propellant. Warheads that use explosives like TNT are very sensitive to heat and shock. These types of warheads are exchanged with Plastic Bonded Explosives which can withstand adverse conditions. Insensitive munition will not detonate under any condition other than its intended mission to destroy a target. The main advantages of insensitive munitions are: increased weapon storage densities, decreased risk of unplanned detonation, reduced vulnerability to distribution of fragments, low vulnerability to mechanical (impact and friction) and shock stimuli and less prone to thermal stimuli including aerodynamic heating. The major thrust to research in the area of insensitive munitions has been provided with the signature of STANAG. This has accelerated R&D in the direction of low vulnerable explosives acronymed as LOVEX, popularly known as PBX which is an explosive material in which explosive powder is bound together in a matrix using small quantities (typically 5-10 per cent by weight) of a synthetic polymer ("plastic"). This type of explosive is already being used in torpedo warheads. Their usage is bound to proliferate to other munitions not only due to low vulnerability but also because of superior structural integrity.

3.6.2 Novel Energetic Materials

Novel Energetic Materials consist of fundamental research to expand and validate physics-based models and experimental techniques to devise chemical formulations that will enable the design of novel insensitive high-energy propellants and explosives with tailored energy release. It supports demonstration of advanced energetic materials with the ability to tune energy release for precision munition and counter-munition applications (e.g., propellants, explosives, thermobaric, multi-purpose warhead, etc.).

These energetic materials may have the potential of providing factors of three to four in increased energy release rate as compared to conventional formulations. Efforts are underway to develop advanced energetic materials to provide a 40 per cent increase in deliverable energy from advanced gun propellant systems and a 20-50 per cent increase in warhead effectiveness (munitions, active protection).

Under normal conditions of use, modern munitions are both effective—they provide an essential military capability—and relatively safe—they are unlikely to explode or burn spontaneously despite the fact that they are composed primarily of hazardous material. Under severe conditions, however, their dangerous nature comes to light. The energetic materials—high explosives, gun propellants, rocket propellants—that are found in munitions of all types are sensitive to heat and to mechanical shock so they may be triggered by fire or by impact with bullets or fragments. Such secondary effects are significant: in the Gulf War, for example, most of the disabling damage to fighting vehicles was found to be caused by their own munition payloads, inadvertently triggered by unwanted stimuli.

A revolutionary propulsion system could dramatically change war fighting and be a critical element in military transformation objectives. This system uses the physics of nano-to-macro-scale explosive burst generation from pre-tensioned meta-stable fluids as simple as water. This new propulsion system can produce energy outputs that are 500 per cent higher than the energy produced using conventional nitrocellulose.

Ultra-high explosive burst generators have applications beyond that of propellants; they could be used as explosives as well as to destroy chemical or biological agents. The use of pre-tensioned fluids will allow generation of energy bursts in a range that is higher than the output of any known explosive or propellant. The concept of a nano-scale burst propellant is to stretch liquids like a spring to desired levels and then use neutrons to safely trigger explosive bursts. The neutrons act as a nano-scale trigger to release energy from critical-sized vapour pockets in liquid. The growth and collapse of the vapour pockets creates intense heat (about 106K to 108K) and localized shock waves.

3.6.3 Composite Propellants

India has developed different composite propellants for aerospace and defence applications and they have been successfully tested. At present it employs six solid propellant strap-on motors, each carrying nine tonnes of propellant. The new version, PSOM-XL, with a length of 13.5 m, has the capacity to carry 12.4 tonnes. The motor developed a peak chamber pressure of 4.16 Mega Pascal and burned for a duration of 58 seconds in the ground test. The performance of the new motor, developed by ISRO's Vikram Sarabhai Space Centre at Thiruvananthapuram, was as per prediction. PSOM-XL will improve the capability of PSLV from the present 1,450 kg to 1,600 kg and will be employed in future PSLV flights including the Chandrayaan programme and the microwave remote sensing satellite, RISAT.

Global demand for longer range and high payload capability

catapulted composite propellants to premier position and relegated double base propellants to a relatively less significant position. Composite propellants are today's workhorse propellants for both defence and space missions. Remarkably tunable properties of Hydroxyl-Terminated Poly Butadiene (HTPB) binder have made it possible to achieve a wide range of mechanical properties needed for free standing and case bonded configurations capable of operating at sub-zero as well as high temperatures. An intense research activity is on all over the world to realize energetic eco-friendly propellants such as Ammonium Di-Nitramide (ADN) which has entered the domain of advanced propellants. Technology of RDX/HMX and polymer-based propellants (Nitramine-based propellants) has also entered the area of gun propellants. This class of propellants designated as Low Vulnerability (LOVA) are on the verge of replacing conventional Double Base (DB) and Triple Base (TB) propellants. The major impetus in this direction has been provided by the catastrophic damage caused due to unplanned initiation of conventional Gun Propellants (GP) during the Gulf war. India has mastered the technology of double base and composite propellants.

3.6.4 Nuclear and Antimatter Propulsion

Nuclear propulsion is a very attractive option for exploration in deep space. Potential fission-based transportation options include high specific power, continuous impulse propulsion systems and bimodal nuclear thermal rockets. Nuclear propulsion offers promising scope in building nuclear rockets which utilize fission energy to heat a reactor core to very high temperatures. Hydrogen gas flowing through the core gets super heated and exits the engine at very high exhaust velocities. The combination of temperature and low molecular weight of hydrogen will produce an engine that is capable of specific impulses above 1000 seconds. Simulation and research studies are focused on building a safe affordable fission engine and its components which include refractory metal modules, heat pipes, high temperature heaters, stainless steel cores, space fuelling, etc.

Matter/antimatter annihilation has the highest energy release per unit mass of any reaction known in Physics and the energy so obtained can be used for space exploration. Antimatter can be thought of as the mirror image of normal matter seen in everyday life. An antiparticle has identical mass to its normal matter twin, but opposite charge and spin. Matter is composed of electrons, protons and neutrons. Each of these particles has an antimatter counterpart referred to as positron, antiproton and antineutron, respectively. For every normal matter particle created in high-energy process such as collisions in particle accelerator, an antiparticle is created. An interesting property of antimatter is its dramatic interaction with its normal matter counterparts. When a particle and an antiparticle come in close contact, they annihilate each other through a series of interactions with the ultimate result that the remaining mass is converted entirely into energy. For every particle known to exist, there is a mirror image twin particle that has its charge, spin, and quantum states reversed from that of normal particles. Stable particles that make up atoms-electrons, protons, and neutrons - have mirror twins called positrons, antiprotons, and antineutrons. Conceptually, these could be combined to form anti-atoms such as anti-hydrogen. Antimatter stores energy at a very high density. Approximately 42mg of antiproton (about 0.6 cc in the form of anti hydrogen) have energy content equal to 750,000 kg of fuel and oxidizer.

When antiprotons annihilate with protons, the products of the annihilation are from three to seven particles called pions. On an average there are three charged pions and two neutral pions. The neutral pions have a very short lifetime and almost immediately convert into two high energy gamma rays. The charged pions have a normal half-life of 28 ns and are moving at 94 per cent of the speed of light. However, their lives are lengthened to 70 ns. These charged pions contain 60 per cent of the annihilation energy. Because of the long lifespan and interaction length of the charged pions that result from the annihilation of antiprotons with protons, it is relatively easy to collect the charged pions in a thrust chamber constructed of magnetic fields and to obtain propulsion from them. Energy in the pions can then be used either to heat a working fluid, such as hydrogen, to produce thrust, or the high speed pions themselves can be directed by a magnetic nozzle to produce thrust. Even after the charged pions decay, they decay into energetic charged muons, which have an even longer lifespan and interaction length for further conversion into thrust.

3.6.5 Large Hadron Collider (Big Bang)

Hadrons are either protons or lead ions of sub-atomic particles and the collider accelerates two beams of particles of the same kind in the opposite direction inside a circular accelerator. Large Hadron

Collider (LHC) is a gigantic scientific  accelerator where these heavy particles like protons moving at high speed are collided to generate high energy, thus recreating the conditions just after the Big Bang. As these charged particles are accelerated in a curved path, the process is called synchrotron radiation. This experiment is expected to reveal the theory behind the formation of the universe, mystery of anti-matter and moving the particles faster than light.

LHC is built in a 27 kilometres tunnel at a mean depth of 100 metres due to geological considerations and at a slight gradient of 1.4 percent. Its depth varies between 175 metres and 50 metres. European Organisation for Nuclear Research (CERN) is a joint venture of twenty member states. India is one among them represented by Department of Atomic Energy.

3.6.6 Nutrino

The neutrino was first postulated in December, 1930 by Wolfgang Pauli to explain the energy spectrum of beta decays, the decay of a neutron into a proton and an electron. Pauli theorized that an undetected particle was carrying away the observed difference between the energy and angular momentum of the initial and final particles. Because of their "ghostly" properties, the first experimental detection of neutrinos had to wait until about 25 years after they were

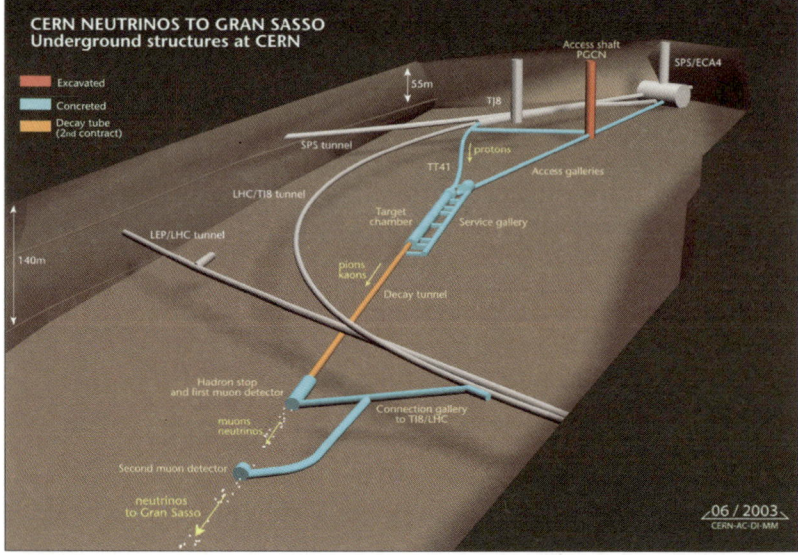

first discussed. In 1956 Clyde Cowan, Frederick Reines, F. B. Harrison, H. W. Kruse, and A. D. McGuire published the article "Detection of the Free Neutrino: a Confirmation" in Science, a result that was rewarded with the 1995 Nobel Prize. Neutrinos are subatomic particles produced by the decay of radioactive elements and are elementary particles that lack an electric charge.

Attempts have been made to attain faster than light speed worldwide. The recent OPERA [Oscillation Project with Emulsion-tRacking Apparatus] neutrino experiment at the underground Gran Sasso Laboratory has measured the velocity of neutrinos from the CERN beam over a baseline of about 732 km with much higher accuracy than previous studies conducted with accelerator neutrinos. The measurement is based on high statistics data taken by OPERA in the years 2009, 2010 and 2011. Dedicated upgrades of the CNGS timing system and of the OPERA detector, as well as a high precision geodesy campaign for the measurement of the neutrino baseline, allowed reaching of comparable systematic and statistical accuracies. In their original experiment scientists fired beams of neutrinos from CERN to the Gran Sasso lab. A beam of light would take just 2.4 milliseconds to cover this distance whereas Nutrinos travelled in 60 nanoseconds (60 billionths of a second) faster than light. Nutrinos were travelling at a speed of 299,798,454 metres per second while the speed of light in vacuum is 299,792,458 metres per second.

India-based Neutrino Observatory (INO)

The India-based Neutrino Observatory (INO) Project is a multi-institutional effort aimed at building a world-class underground laboratory with a rock cover of approx.1200 m for non-accelerator based high energy and nuclear physics research in India.

The project includes:

a) construction of an underground laboratory and associated surface facilities at Pottipuram in Bodi West hills of Theni District of Tamil Nadu,

b) construction of an Iron Calorimeter (ICAL) detector for studying neutrinos, consisting of 50000 tons of magnetized iron plates arranged in stacks with gaps in between where Resistive Plate Chambers (RPCs) would be inserted as active detectors, the total number of 2m * 2m RPCs being around 29000, and

c) setting up of National Centre for High Energy Physics at Theni near Madurai, for the operation and maintenance of underground laboratory, human resource development and detector R&D along with its applications. The underground laboratory, consisting of a large cavern of size 132m * 26m * 20m and several smaller caverns, will be accessed by a 2100 m long and 7.5 m wide tunnel.

The initial goal of INO is to study neutrinos. Neutrinos are fundamental particles belonging to the lepton family. They come in three types, one associated with electrons and the others with their heavier cousins the muon and the Tau. According to standard model of particle physics, they are massless. However recent experiments indicate that these charge-neutral fundamental particles have finite but small mass which is unknown. They oscillate between flavours as they propagate. Determination of neutrino masses and mixing parameters is one of the most important open problems in physics today. The ICAL detector is designed to address some of these key open problems in a unique way. Over the years this underground facility is expected to develop into a full-fledged underground science laboratory for other studies in physics, biology, geology, hydrology etc.

3.7 NUCLEAR ENERGY

Energy released during a nuclear reaction as a result of fission or fusion is termed as nuclear energy also called atomic energy. The power from nuclear energy was first discovered in 1896 by a French physicist named Henri Becquerel who saw that some photographic plates like X-ray plates which were stored near uranium turned dark. The first nuclear power station was established in 1956 in Cumbria, England. Nuclear energy is used in many other capacities and suffices about 16 per cent of the Earth's energy requirements and has many advantages such as: it is more affordable in producing energy than energy produced through coal, does not use as much fuel in the process, produces less waste and is a clean source of energy.

As carbon-based fuels grow increasingly scarce in the face of ever-growing demand, new and more sustainable sources of energy will be necessary to meet global energy needs. By 2050, an expected rise in global population from six billion to nine billion and better living standards could lead to a two to threefold increase in energy consumption. No single technology will fulfill this demand. Though each technology has got its own strengths and weaknesses, a mix of power sources will be needed to meet the challenges of energy security, sustainable development and environmental protection.

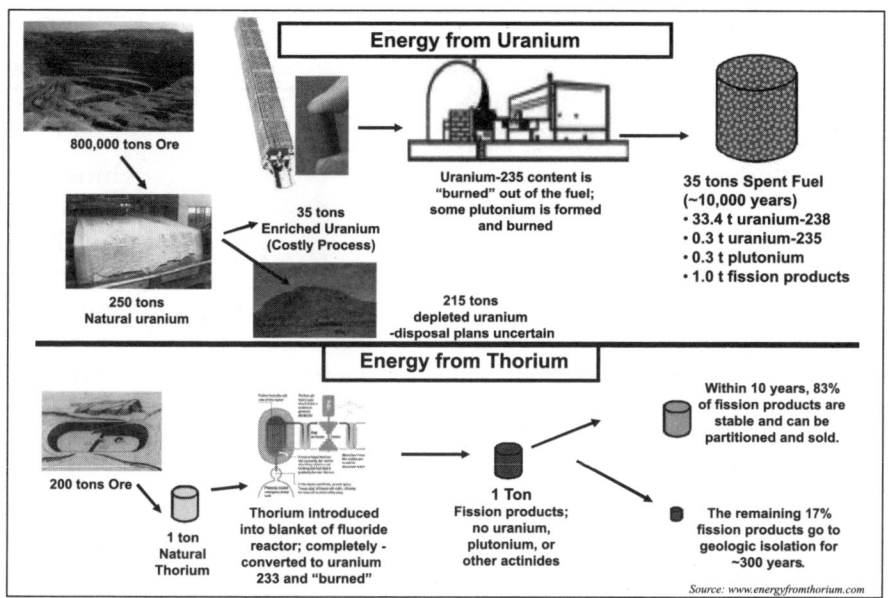

Figure: 3.45: Uranium Fuel Cycle vs. Thorium

Energy supply options comprise fossil fuels, nuclear fission and fusion. Nuclear fission continues to be the major contributor for electricity generation but the question of safety and disposal of spent fuel might restrict the generation of power. Supplies from the renewable sources are heavily dependent on the environmental conditions and the challenge for technology is the storage of energy.

Every single atom in the universe carries an unimaginably powerful battery within its heart called the nucleus. This form of energy, often called Type-1 fuel, is hundreds of thousands, if not millions, of times more powerful than the conventional Type-0 fuels, which are basically dead plants and animals existing in the form of coal, petroleum, natural gas and other forms of fossil fuel. To put things in perspective, imagine a kilometre-long train, with about 50 freight bogies, all fully laden with the most typical fossil fuel about 10,000 tonnes of coal. According to *world-nuclear.org,* the same amount of energy can be generated by 500 kg of Type-1 fuel, naturally occurring Uranium, enough to barely fill the boot of a small car. When the technology is fully realised, one can do even better with naturally occurring Thorium, in which case the material required would be much less, about 62.5 kg, or even less according to an estimate by A Canon Bryan. (Note: 500 kg of naturally occurring Uranium would contain about 3.5 kg of Uranium-235 fuel.)

3.7.1 Energy and Economy

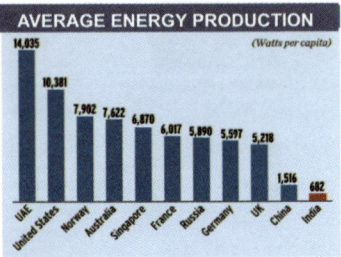

Energy is the most fundamental requirement of every society or nation as it progresses through the ladder of development. Of course, once it reaches a relative degree of development, the energy demand becomes more stable. There is a distinct and categorical correlation between the energy consumption and income of a nation—each reinforcing the other. Look around you: every step into progress comes with an addition in demand for energy—cars, ships and aircraft to move, hospitals to give quality healthcare, education, as it follows the model of e-connectivity, production of more and better goods, irrigation for better farming. In fact, every element of our lives is increasingly going to become energy-intensive —that is a necessary prerequisite for development. This is clearly reflected in the average energy consumption per person across nations—for instance, an average American consumes more than 15 times the energy consumed by an average Indian.

Today, India needs to rebuild its economy with higher GDP growth rate from industries. Our focus for this decade is therefore the development of key infrastructure and industrial output for the upliftment of the 600,000 villages where 750 million people live. In order to achieve the growth of economy on expected lines, it is predicted that the total electricity demand will grow from the current 150,000 MW to at least over 950,000 MW by the year 2030 which will still be less than one-fourth of the current U.S. per capita energy need.

The study indicates that most of the prosperous nations are extracting about 30-40 per cent of power from nuclear power and it constitutes a significant part of their clean energy portfolio, reducing the burden of combating climate change and the health hazards associated with pollution. India is generating a mere 4,780 MW of nuclear power from a total of about 207,000 MW of electricity generation, most of it coming from coal. Besides the billions spent on importing coal or oil, we are also importing millions of tonnes of CO_2 and other greenhouse gases which are a hazard to the environment and human health. Hence it is essential to go a big way in the generation of nuclear energy.

Fission Process with Uranium

Uranium fuel cycle vs. Thorium for generation of 1000 MW of electricity for one year is shown in figure (Figure 3.45). 800000 tonnes ore which is converted into approximately 250 tons of uranium i.e. only 0.7 per cent of ore contains the radio isotope. Approximately 35 Tonnes of enriched Uranium and 215 Tonnes of depleted Uranium are generated through a costly process. This enriched Uranium burned out of the fuel produces 1000 MW of electricity for one year. In this process, it is left with approx 35 tonnes of spent fuel which are radioactive materials with half life period of approx 10,000 years.

3.7.2 Thorium: A Futuristic Nuclear Fuel

India is blessed with a rare, and very important, nuclear fuel of the future – Thorium. More than 30 per cent of the world stock of Thorium is in India. We should utilise this opportunity to emerge as the energy capital of the world which results in the emergence of India as the leading economy of the world. India has the potential to be the first nation to realise the dream of a fossil fuel-free nation which will relieve the nation of about $100 billion annually that we spend in importing petroleum and coal. Hence, Thorium is no doubt, a fuel of the future, especially for India.

A lesser-known member among radioactive materials is Thorium. Perhaps it is the best solution possible in the future and would be technologically and commercially the best option in another two decades. Thorium, the 90th element in the Periodic Table, is slightly lighter than Uranium. Thorium is far more abundant, about four times more abundant, than the traditional nuclear fuel, Uranium and occurs in far purer form too. It is believed that the amount of energy in Thorium reserves on earth is more than the combined total energy in left petroleum, coal, other fossil fuels and Uranium, all put together. The figures as revealed in the IAEA Report (2005) on Thorium fuels, indicate that India has the largest reserves of Thorium in the world, with over 650,000 Tonnes. Comparatively we have barely 1 per cent of the world Uranium deposits which is currently being put to effective use through our opting for closed fuel cycle technology. Thorium has many other advantages. It is estimated that Thorium may be able to generate (through uranium-233 producible from it) eight times the amount of energy per unit mass compared to (natural) Uranium. The much debated waste generation is also at a relative advantage for Thorium. It produces waste that is relatively less toxic due to the absence of minor actinides (associated with uranium). The energy

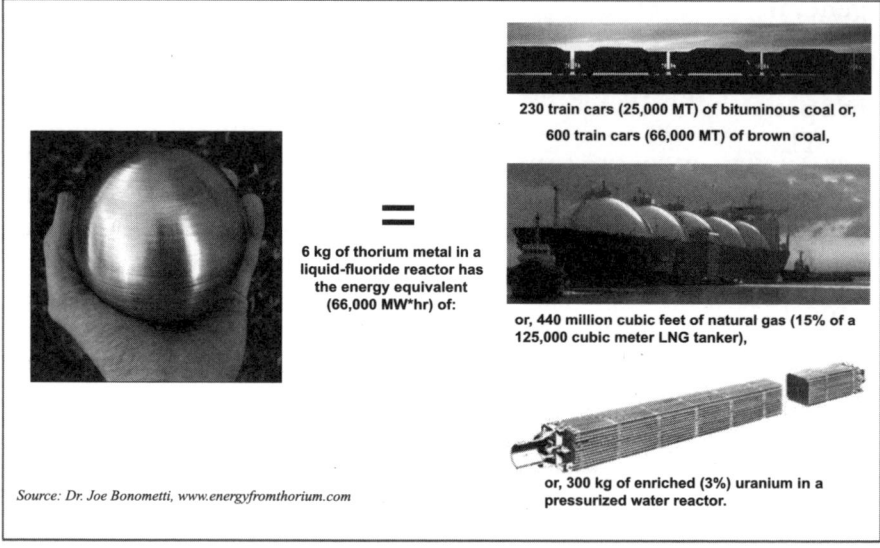

Source: Dr. Joe Bonometti, www.energyfromthorium.com

Fig. 3.46: Energy Generation Comparison

	Uranium LWR (light water reactor, high pressure low temp)	Thorium LFTR (liquid fluoride thorium reactor, low pressure high temp)
Plant Safety	Good (high pressure)	Very Good (low pressure, passive containment)
Burn Existing Nuclear Waste	Limited	Yes
Radioactive Waste Volume (relative)	1	1/30th
Waste Storage Requirements	10,000+ yrs.	~300 yrs.
Produce Weapon Suitable Fuel	Yes	No
High Value By -Products	Limited	Extensive
Fuel Burning Efficiency	<1%	>95%
Fuel Mining Waste Vol. (relative)	1000	1
Fuel Reserves (relative)	1	≻1000
Fuel Type - Fuel Fabrication/Qualification	Solid Expensive/Long	Liquid Cheap/Short
Plant Cost (relative)	1 (high pressure)	<1 (low pressure)
Plant Thermal Efficiency	~35% (low temp)	~50% (high temp)
Cooling Requirements	Water	Water or Air

Source: http://www.energyfromthorium.com/ppt/thoriumEnergyGeneration.ppt

Fig. 3.47: Comparison Between Uranium vs. Thorium Based Nuclear Power

generation comparison of 6 kg of thorium metal in a liquid-fluoride reactor has the energy equivalent of 25000 MT of bituminous coal or 66,000 MT of brown coal or 300 kg of enriched (3 per cent) uranium in a pressurized water reactor (Figure 3.46).

Thorium Fuel Cycle (Fission Power): In typical energy generation, 200 Tonnes of ore is converted into 1 ton of natural thorium which is introduced into a blanket of fluoride reactor; completely converted to uranium-233 and "burned", producing 1000 MW of electricity for one year as explained in Figure 3.52. In this process, it is left with 1 Ton Fission products that contain no uranium, plutonium, or other actinides. Within 10 years, 83 per cent of fission products are stable and can be partitioned and sold, and the remaining 17 per cent fission products go to geologic isolation for ~300 years. The separated fission products out of liquid fluoride thorium reactor will have many valuable by-products such as Strontium-90 for radioisotope power, Cesium-137 for medical sterilization, Rhodium, Ruthenium as stable rare-earths, Technetium-99 as catalyst, Molybdenum-99 for medical diagnostics, Iodine-131 for cancer treatment, Xenon for ion engines. These products may be as important as electricity production. The relative comparison between the Uranium based and Thorium based nuclear power on various factors is given in the figure (Figure 3.47).

Now, being the largest owner of Thorium, and also being amongst the nations which will see the highest surge in power demand with its growth, the onus is on India to pursue its existing nuclear programme with a special focus on research and development of Thorium route as a long term sustainable option. For this purpose, it is imperative to continue to implement the current Indian plan of making use of the uranium and plutonium based fuel cycle technologies as well as irradiate larger amounts of thorium in fast reactors for breeding uranium-233 fuel as it undergoes a graduation to the Thorium based plants. It is noteworthy that the Indian plan for an advanced heavy water reactor (AHWR) is an important step to launch early commencement of thorium utilization in India, while considerable further efforts to use thorium in both thermal and fast reactors would be essential for harnessing sustainable energy from thorium-generated uranium-233. Various technologies for Thorium based plants are already being developed and deployed on test basis across the world including India. These include first breeding it to fissile Uranium 233 isotope in conventional reactors or through revived interest in the technologies like the Molten Salt Reactors (MSR) which use salts to trap the fissile material and do not react with air or burn in air or water. In this technology, the operational pressure is near to the ordinary atmospheric pressure hence the cost of construction is low and there is no risk of pressure explosion.

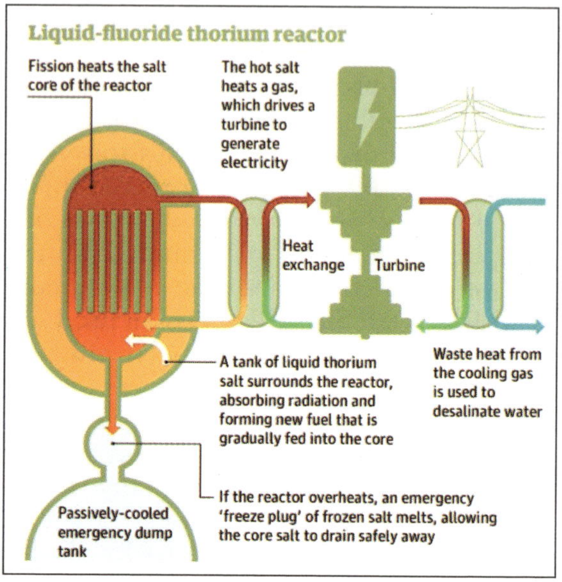

Fig. 3.48: Liquid Fluoride Thorium Reactor (LFTR)

The Liquid Fluoride Thorium Reactor (LFTR) also called the Molten Salt Reactor (MSR) is presently the cleanest, safest, and most efficient nuclear power reactor ever developed. It operates at high temperatures and at ambient air pressures, and can be scaled to any size required. The typical functioning of the Liquid Fluoride Thorium Reactor (MSR/LFTR) is shown in Figure 3.48.

Advantages include:

1) There is no pressure in the reactor system so that it cannot explode – unlike traditional nuclear reactors which operate as a high pressure steam boiler.
2) The fuel fabrication is easier. The thorium fuel does not need to be shaped into pellets; it is dissolved into a liquid.
3) The reactor can have fuel added and waste removed at any time online with normal operations.
4) There are no weapons-grade materials involved.
5) Above all, thorium is abundant, and 97 per cent of it gets converted to power in the reaction, (with uranium only 5 per cent gets used).

3.7.3 Fusion Technology

Nuclear fusion is the process by which two or more atomic nuclei fuse together to form a single heavier nucleus. This is usually accompanied by the release or absorption of large quantities of energy. The Fusion process powers active stars. The occurrence of fusion of hydrogen atoms at the core of the sun at a temperature of about 10 million degrees, is the natural phenomena of nuclear fusion generating more energy. The same fusion power may be used for electricity generation.

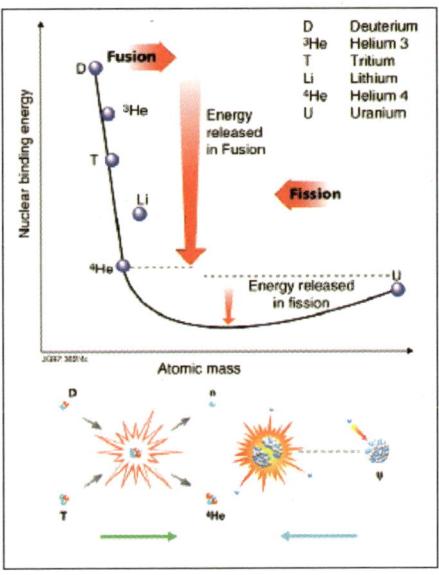

Fusion power has the potential to provide sufficient energy to meet an increasing demand with a relatively small impact on the environment. Commercial fusion power would guarantee the safe, CO_2, pro environmental energy production with abundant and widely available fuels. The fuels Deuterium (D) and Tritium (T) are the easiest fuels for fusion reaction. Deuterium is extracted from water and tritium is produced from Lithium which is abundant in earth's crust and in sea.

Scientists and engineers are developing technology to use this Fusion process in tomorrow's power stations. To achieve high enough fusion reaction rates to make fusion useful as an energy source, the fuel (deuterium and tritium) is heated to temperatures over 100 million degrees Celsius. At these temperatures the fuel becomes plasma. This extremely hot plasma is very thin and fragile, a million times less dense than air. The plasma is confined using strong magnetic fields that create a sort of fence preventing the particles to touch and damage the walls of the device. These fuse to produce helium and high-speed neutrons, releasing a high amount of energy per reaction. A commercial fusion power station will use the energy carried by the neutrons to generate electricity. The neutrons will be slowed down by a blanket of denser material surrounding the machine, and the heat produced would be converted into steam to drive turbines and put power on to the grid.

There are a number of merits of the fusion. This would be a large source of energy for which the basic fuels are available in sea in a large quantity. During the process of fusion energy generation, the only byproducts are small amounts of helium-which is an inert gas that will not add to atmospheric pollution and hence does not emit any greenhouse effect and hence has a very low impact on the environment. Since the basic fuel for fusion energy is two types of Hydrogen (D and T) which are abundant fuel supplies that will last for millions of years. There will be no long-lasting radioactive waste. Only plant components become radioactive and these will be safe to recycle or dispose of conventionally within hundred years. Fusion power stations would be inherently safe as there is no possibility of meltdown. These reactors work like a gas burner. The moment supply of fuel is closed, the reaction ends. The future 1GWe fusion power plant would be able to operate for one year with 3000 m^3 of water and 10 tons of Li ore.

3.7.4 ITER Project – The Road to Fusion Power

ITER (An acronym of International Thermonuclear Experimental Reactor) is an international nuclear fusion research and engineering project which is currently building the world's largest and most advanced experimental nuclear fusion reactor. The ITER project aims to make the long-awaited transition from experimental studies of plasma physics to full-scale electricity-producing fusion power plants.

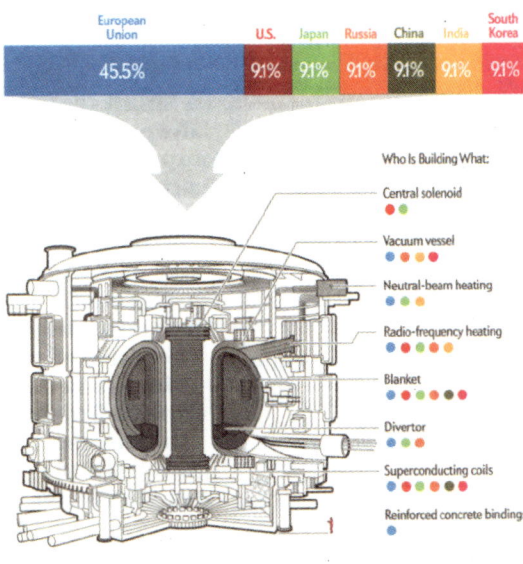

The project is funded and run by seven member entities — the European Union (EU), India, Japan, China, Russia, South Korea and the United States. ITER will produce about 500 MW output power in nominal operation with an input heating power of 50 MW and hence the power amplification is 10. ITER is based on the doughnet-shaped "Tokamak" Concept surrounded by coils that produce intense magnetic field. The first plasma is expected to be produced in five years time. According to the ITER Project, after the successful construction of the first electricity generating plant, the reliable and commercially available electrical power from fusion would be available around 2045. The longterm aim of fusion is to create power station prototypes taking into account the operational safety, environmental compatibility and economic viability. It is required to concentrate on technology development and concept improvement for realisation of commercial electricity producing reactor.

The power of the nucleus is mighty and the future of humanity lies in harnessing it in a safe and efficient manner. In the years to come, it will fuel not only our earth-based needs but also our space missions and perhaps even our civilisation's reach to other planets for habitation. Our current nuclear projects will expand into better and safer materials, like Thorium, and later on, into better reactions like fusion, which once completely developed, will be able to generate hundred times more power than current fission methods. Affordable, clean and abundant energy provided by nuclear sources is our gateway to a future that is healthy, learned and connected, a future that will span deep into space and cross the boundaries of current human imagination.

3.8 SPACE TECHNOLOGY

Space Visionaries

Acharya Bhardwaj (India) designed three categories of flying machines in 800 BC for travel between planets. Konstantin Tsiolkovsky (USSR) formulated the first theoretical framework for space exploration in 1903, and gave a concept of multistage rockets and envisioned man landing on Mars. Herman Oberth (Germany) formulated the idea of manned flights  searching for life wherever possible. Robert Goddard (USA) tested liquid propellant rocket and was the first to propose a rocket to the moon. Wernher von Braun in Germany developed the first guided missile V2 in October 1942 and introduced it in the Second World War. He also developed the first long range ballistic missile "Red Stone" for USA. His Saturn V put man on the moon. In USSR, Sergei Korolev developed the first Soviet liquid propellant rockets, eventually the first ICBM. Later, this ICBM was converted into a launch vehicle and placed Sputnik in orbit.

Firsts in Space

The Soviet Union launched the world's first artificial earth satellite

SPUTNIK into orbit on 04th October 1957, marking the beginning of space ventures. USA soon followed in 1958 and the race for conquering the space started between the two giant nations. With the historical launch of Vostok-1 on 12 April 1961, Yuri Gagarin became the first man to go to space, see the  earth from space and return.

 Many scientific experiments were carried out by NASA to study the moon and different planets. Neil Armstrong landed on the moon's surface giving the message of a giant leap for mankind and heralded a new dimension to the space programme. Photographs and samples taken from the moon and biological aspects of living will definitely lead to a new era in space in the years to come. International Space Station is already in the orbit for scientific experiments. Space is also used for military purposes for communication and remote sensing. With huge opportunities knocking at mankind's door, space is becoming an enterprise. India planned systematic growth in Science and Technology after independence, and in the early 1960s the seed of space research was sown. Rakesh Sharma was the first Indian to go to space in a Soviet capsule to see the earth from space in April 1984.

World's First Rocket – Made in India

The first 'ballistic weapons' probably were rocks that caveman hurled at each other. These 'missiles' were followed by sticks fitted with pointed stoneheads to make spears and later by wood and 'string' devices that propelled smaller wooden shafts through the air. Chinese used fire arrows. In the eighteenth century, an interesting innovation happened in India by using, for the first time, a war rocket. Started by Hyder Ali, his son Tipu Sultan used the world's first war rocket in the Srirangapatna war in 1792, launched in huge numbers against the British cavalry, and defeated them.

These were extremely effective in battle and inflicted heavy losses on British forces. Fighting the British East India Company, Tipu Sultan's army used a variety of rockets in a supporting role. It was the world's first use of rockets in modern wars which was later developed further by the British. In the second Anglo-Mysore war, at the Battle of Pollilur (10 September 1780), an entire British detachment led by Colonel Baillie was destroyed resulting in 3,820 soldiers being taken prisoner including Colonel Bailli. At the Battle of Srirangapatana in 1792, Mysorean soldiers launched a barrage of rockets against British troops followed by an assault by 36,000 men. Later, in the battle of Srirangapatana during the fourth Anglo-Mysore war in April 1799 British forces retreated from the battlefield when attacked by rockets and musket fire from Tipu Sultan's army. After the capture of the Mysore kingdom, the rocket production plants were destroyed by the British army. Thousands of stored rockets were taken to Britain and stored at Royal Artillery Museum, Woolwich and the British Royal Arsenal.

Tipu's rockets of that period were much more advanced than any known artillery weapon used in the war at that time (Figure 3.49). Tipu Sultan wrote a military manual called "Fathul Mujahidin" in which two hundred trained rocketmen were assigned to each artillery brigade called Kushoon. There were twenty four Kushoons forming the infantry battalion. The rocket men were trained to launch these rockets from wheeled rocket launchers capable of launching five to ten rockets simultaneously. The launch angle was calculated from the diameter of the cylinder and the distance of the target. Rockets would be of different size. The typical rocket consisted of a tube of 60 mm diameter and 250 mm long with 2 kg gun powder, tied to a sword which could reach a range of 1.0 to 1.5 km. The metal cylinder used was made of hammered soft iron. The use of iron increased bursting pressure which permitted the propellant to be packed to greater densities thus giving the rocket a higher thrust and range. Thanks to

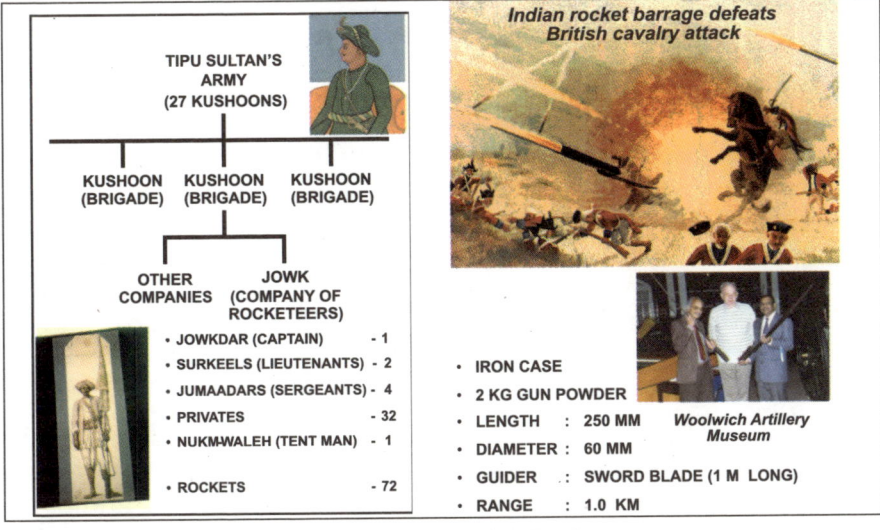

Fig. 3.49: World's First Rocket

the British they kept Tipu's rocket in the Royal Artillery Museum, Woolwich, London, still reminding us of the extraordinary skill of Indians.

In 1801, Sir William Congreve set a research and development effort at the Arsenals Laboratory to understand and re-engineer Tipu's rocket. He made rockets of the same design with increased range up to 3 km. The British used these rockets for war with Napoleon at Waterloo in 1809. Napoleon immediately offered a reward to any French inventor who could replicate these rockets for the French force. Men succeeded during Napoleon time. The British installed these rockets on onboard ships and used against the American defence at Fort Henry in 1812. The "Rockets red glare" is enshrined in the American national anthem in its fifth line. This is how the Indian war rocket developed and produced in India travelled to many countries and was re-engineered. That is why in the Royal Artillery Museum in Woolwich, London is written that, "The rocket technology came from the east to west".

3.8.1 Space Technology in India, after Independence

The Indian space programme started way back in 1961 when the Government of India entrusted the study of space research to the Department of Atomic Energy (DAE) under the able leadership of Dr. Homi J. Bhabha. In the very next year, DAE set up the Indian National Committee on Space Research (INCOSPAR) with Prof. Vikram A Sarabhai as the Chairman. The Indian space programme was driven by the vision of Dr. Vikram Sarabhai, the father of Indian Space Programme.

Dr. Sarabhai quoted:

"There are some who question the relevance of space activities in a developing nation. To us, there is no ambiguity of purpose. We do not have the fantasy of competing with the economically advanced nations in the exploration of the moon or the planets or manned space-flight. But we are convinced that if we are to play a meaningful role nationally, and in the community of nations, we must be second to none in the application of advanced technologies to the real problems of man and society."

Both Dr. Homi Bhabha and Prof. Vikram Sarabhai were looking for a site to establish space research station in the equatorial region. These two great scientists visited a number of places. Thumba in Kerala was selected by the scientific community for space research as the magnetic equator of the earth passes through this place, and it was ideally suited for ionospheric research in upper atmosphere apart from the study of atmospheric structure.

The Thumba Equatorial Rocket Launching Station (TERLS) took shape and rocket activities started in the Church of Reverend Father Peter Bernard Pereira who was magnanimous to give the Church and his cottage for space research. The space programme was started with sounding rockets for scientific experiments and Thumba was dedicated to the United Nations for launching sounding rockets of other countries. The first sounding rocket (Nike-Apache) was launched from TERLS on 21 November 1963.

Dr. Sarabhai gave a herculean thrust to space research and technology build-up by bringing the best Indian scientists working in Indian and foreign laboratories and assembled the rank holders of universities. His passion for leadership gave a tremendous momentum to space activities in India.

The indigenously developed first Indian rocket called Rohini-75 was launched on 20 November 1967 from Thumba. Encouraged by the success, the engineers made a two-stage rocket called Centaur and started its regular production in 1971. The Rohini series expanded. The diameter of the rockets steadily increased along with their sophistication. Rohini-100, RH-300 and RH-560 were soon launched, both for carrying payloads to study the atmosphere and for testing new subsystems for bigger launch

RH-75

vehicles. The progress made in sounding rockets led to the design and development of a Satellite Launch Vehicle (SLV-3).

Our first attempt to launch SLV E-01 on 10 August 1979 was not successful due to the failure of a component (solenoid valve) in the

second stage control system. After analysis and rectification procedures, the second experimental launch of SLV-3 (SLV E-02) on July 18, 1980 from SHAR Centre Sriharikota achieved the mission objective of injecting Rohini satellite, RS-1 in the low earth orbit. India became the seventh nation in the world to have achieved satellite launching capability after Soviet Union (1957), USA (1958), France (1965), Japan (1970), China (1970), UK

(1971). Encouraged by the success of SLV-3, ISRO visualised the PSLV configuration for placing 1000 kg remote sensing satellite in the polar orbit, while going ahead with SLV-3 continuation and Augmented Satellite Launch Vehicles for increased payload.

PSLV Configuration Design

Configuration of PSLV as an operational launch vehicle for ISRO and developing critical technologies became a challenge in the early 80s for ISRO. Prof. Satish Dhawan, the institution builder, a great thinker and a dynamic leader who envisioned the long-term applications of space research and decided to evolve the PSLV and GSLV configurations with sound technological base to give sustenance to ISRO for many years to come (Figure 3.50).

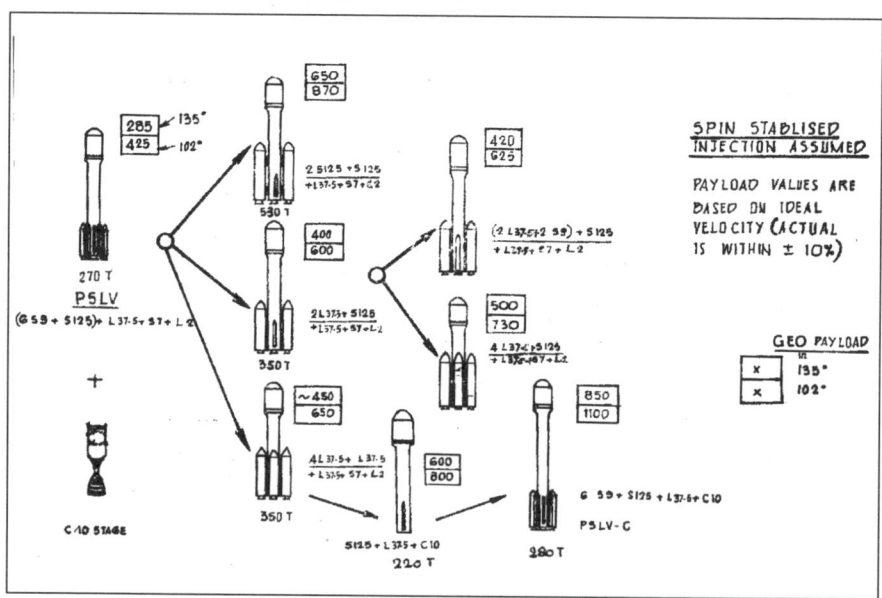

Fig. 3.50: Prof. Satish Dhawan's Launch Vehicle Profile

In 1981, the scientists of Vikram Sarabhai Space Centre, Thiruvananthapuram, with help from other ISRO centres evolved a configuration of PSLV core vehicle with two large strap-on boosters, the PSLV weighing about 400 tonnes at take-off. Prof. Dhawan wanted to study an alternative and simple configuration which can

become workhorse with high reliability. The study at ISRO
Headquarters that our team undertook led to mission analysis,
technology choices and feasibility studies for the optimal
configuration. Our team at ISRO Hqrs. designed several options
including a unique core vehicle with an advanced solid propellant
booster, using first stage rockets of SLV 3 as strap-ons. This brought
the PSLV weight down to only 275 tonnes at take-off. Prof. Satish
Dhawan, himself a foremost aerodynamic engineer with sound
background of mathematics and system engineering illustrated his
ideas on the blackboard and asked us to do more home work. We
studied the performance effectiveness of different launch vehicles
options, cost effectiveness and growth potential with cryogenic upper
stage as GSLV, and the possibility of launching due-east geosynchronous missions overcoming range safety issues. Detailed
examination and debate, taking the long term plans into account,
took place and finally ISRO chose the PSLV configuration as
proposed by our launch vehicle team at the Headquarters. This
unique configuration which is being used with stage-by-stage
upgradation put ISRO in a strong footing with reliable workhorse
PSLV.

Prof. Dhawan's technology mastery, Dr. Brahmprakash's
conviction to use a large booster with Maraging Steel (M250) and our

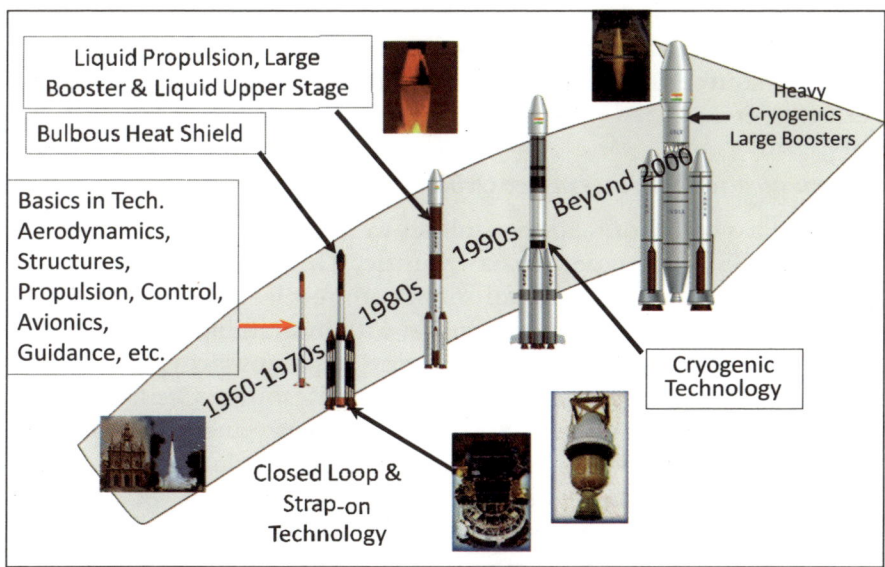

Fig. 3.51: Evolution of Indian Satellite Launch Vehicles and Technologies

(Image Source: ISRO)

team's round the clock mission studies and analysis contributed to the decision on the ever reliable PSLV configuration.

The two operational launch vehicles, Polar Satellite Launch Vehicle (PSLV) and Geosynchronous Satellite Launch Vehicle (GSLV) have continued to provide valuable launch services to our nation establishing self-reliance with launch and satellite capability and also offering commercial launch services to other countries.

3.8.2 Space Technology Applications

Prof. Vikram Sarabhai pioneered India's space programme and unfurled the socio-economic application oriented space mission for India in 1970, which in the last four decades has been touching the lives of billions of people in India in several ways. Today, India with multiple space research centres, supported by many academic institutions and more than 500 industries, has the capability to build any type of satellite launch vehicle to place remote sensing, communication and meteorology satellites in different orbits. Space application has become a part of our daily life.

One of the foremost advantages of space technology is that it can provide a change in the life of the common man. With its reach and potential benefits, people from all over the country are being benefited by space applications. In the past few decades satellite-based communication and remote sensing technologies have provided services related to education, healthcare, weather, land and water resource management, mitigation of impact of natural disasters, etc.

Communication Satellites in Geo Orbit

INSAT class communication satellites were injected in geostationary orbit by GSLV, Ariane, Space Shuttle, Delta launch vehicles at different times for the applications of communication, broadcasting, weather prediction, disaster warning, with availability of twenty four hours in a day. These communication satellites provide valuable support for national development. Village Resource Centres (VRC) is a collaborative programme with NGOs/trusts and governmental agencies to provide a variety of space-based products and services such as tele-education, telemedicine, information on natural resources, interactive advisories on agriculture, fisheries, land and water resources management, livestock management; interactive vocational training towards livelihood support, etc. More than 473

VRCs have been set up in various states of the country and many more are in the pipeline. EDUSAT is India's first satellite dedicated exclusively to educational services. The satellite is specially configured to relay through the audio-visual medium, employing a multi-media multi-centric system to create interactive classrooms. EDUSAT is already providing a wide range of educational delivery modes like one-way TV broadcast, interactive TV, video conferencing, computer conferencing, web-based instructions, etc. Many educational institutions are benefitting enormously by EDUSAT applications. In all, there are 74 networks connecting 55050 terminals [4,050 Interactive and 51,000 receive only] covering 23 States and 3 Union Territories in the country benefitting 15 million students every year (Figure 3.52).

Telemedicine is another important initiative to use space technology for societal benefits. It consists of telemedicine systems-software, hardware and communication equipment aided by satellite bandwidth. Specialty hospitals and state governments provide necessary infrastructure, manpower and maintenance of systems. Currently there are 382 hospitals in the telemedicine network including 306 in remote and rural areas, 60 super-specialty hospitals in major cities and 16 mobile clinics. The other benefits of electronic

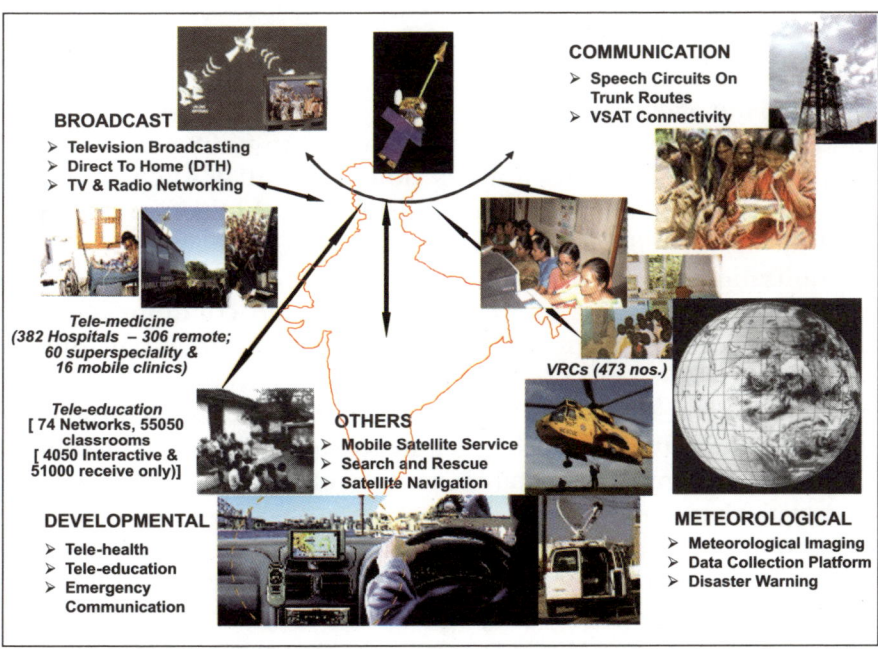

Fig. 3.52: Electronic Connectivity through Space Technology

connectivity are Village Internet Kiosks, Tele-training in farming, e-banking, e-market, e-governance, etc.

3.8.3 Geospatial Technologies for Sustainable Development

India has deployed successfully more than twelve remote sensing satellites for applications such as wet land and wasteland mapping, forest cover, urban planning, ocean resources analysis, crop information, border area monitoring etc. The data generated is processed and stored as database and made available to millions of people in India through organized state level resource management system of ISRO. In addition, the remote sensing data is being shared by users from many countries, having multiple ground stations throughout the world. Remote sensing satellite data is able to quantify and provide macro information in multiple areas. Wasteland of 40 million hectares has been reported presently available whereas the agricultural land used in the country is around 170 million hectares. One of the very important contributions of earth observation is the evolution of a ground water map which has been recently released. This single action has provided availability of water and many citizens have benefitted. It is reported by users that water availability is confirmed in 93 per cent cases while using earth observation data.

Agriculture

About 38 out of every 100 workers in the world are into agriculture. In the Least Developed Nation this ratio goes up to 68 per cent. Our urgent challenge of this hour is how Geospatial Knowledge can enhance the potential of this workforce. Can we think of:

a) Delivering better crop productivity by enhancing the spatial utilization of the farmland, characterizing the soil content, agro climatic variations and changing weather conditions of the farmland? Can we manage the information from end-to-end for the crop cycle?

b) Mapping the water content of all water bodies, their silting status and their potential to bring welfare to the farmlands. Can this be a verified, legally acceptable source of information which is regularly updated?

c) Better pest management and weed management by using GIS application to detect and map the spread of pests in the region and highlighting the vulnerable areas for urgent action.

d) One of the major challenges today is post-harvest management.

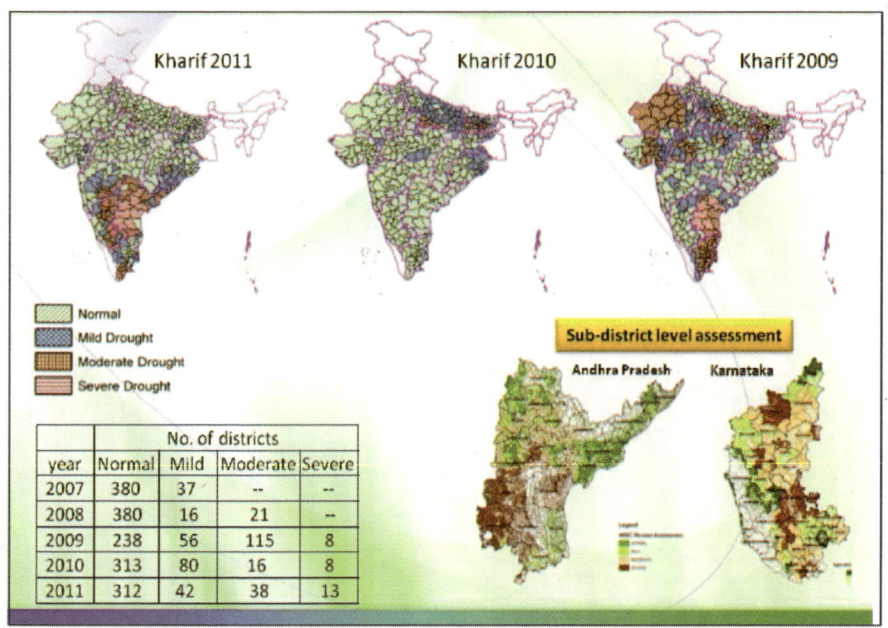

year	No. of districts			
	Normal	Mild	Moderate	Severe
2007	380	37	--	--
2008	380	16	21	--
2009	238	56	115	8
2010	313	80	16	8
2011	312	42	38	13

Fig. 3.53: Space Based Agricultural Drought Assessment

This includes cold storages, silos and markets for food and other farm produce. Geospatial community link can provide information over IVRS in local language for the farmers. This will greatly benefit the agricultural community. This geospatial community link can be undertaken as a social enterprise.

e) All this information be made available in a farmer friendly manner, in his language and delivered over a mobile phone. India alone has about 200 million mobile phone users in rural areas, many of whom are farmers and fisherman. Thus the communication network can empower the farmers and fishermen.

Fishing

Indian National Centre for Ocean Information Services is provides information to fishermen about potential fishing zones. Information so provided is beneficial to the fishermen leading to a big catch. The Indian national centre for ocean information services is using the satellite data of the ocean temperature and colour (chlorophyll), Potential Fishing Zone twice a week (Figure 3.54).

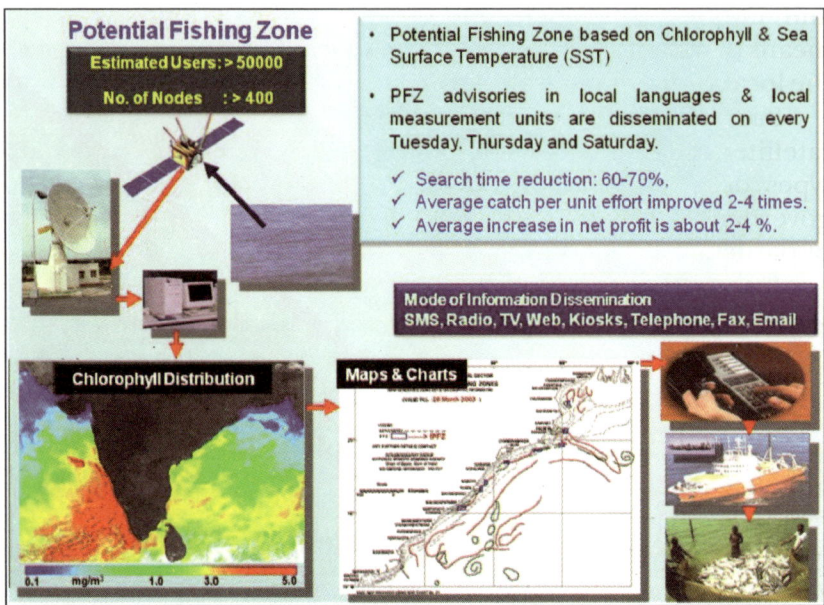

Fig. 3.54: Geospatial Applications in Fisheries for Societal Benefit

Based on this colour and temperature data, they are able to establish potential fishing zone in different parts of the Indian coast line. This information is sent to each landing station in the coast which details about where the potential fishing zone, its distance from the coast, the baring in which one has to go and the depth of the zone is displayed for the information of the fishermen. The data collected at the landing station may be communicated back to the fishermen inside the sea through SMS or other means of communication in the mode of radio waves or FM or through satellite communication. This information becomes a vital tool for the fishermen in the region.

Water, Land and Mobile Resource

The world's water resources are facing potential threats from various forms of water mismanagement either man made or due to natural calamity. Global Climate change has changed dynamics of the environment resulting in flood and drought, in changed dimensions and seasons, which alter the geographic areas in land, river and coastal regions. Our water reserves in deep water table are being drawn out at alarming rates without being replenished. The quality of our water is being contaminated by pollutants, sediments and sewage. Our river ways are becoming clogged with sediments due to erosion. These aspects need continuous observation for preservation, upkeep and improvement. Certainly Geospatial technology provides the means to monitor, measure, model and manage these resources from the local to the global scale. It can be done nationally and globally.

Space faring nations in the world have deployed many ocean satellites; India herself has a series of ocean satellites to explore all types of ocean wealth. These ocean satellites will facilitate geo-governance through the creation of commonly accepted public policy by participating nations.

Interdisciplinary data collection in coastal upwelling regions, seafloor spreading centers, where tropical storms and hurricanes form, where oil spills occur, etc will address many challenges, and the data will be collected from various platforms, instruments, at different study sites, at different scales and resolutions within these study sites. So we are going to continue ways to organize, mine and translate between data which comes from Meta data. This will allow us to maintain and exchange data and information over large distances and long time scales. Bhuvan provides a range of services enabling visualization of various thematic data generated from various satellite resources (Figure 3.55).

GIS and remote sensing are indeed "enabling technologies" for marine science. We know the adage "location, location, location." But

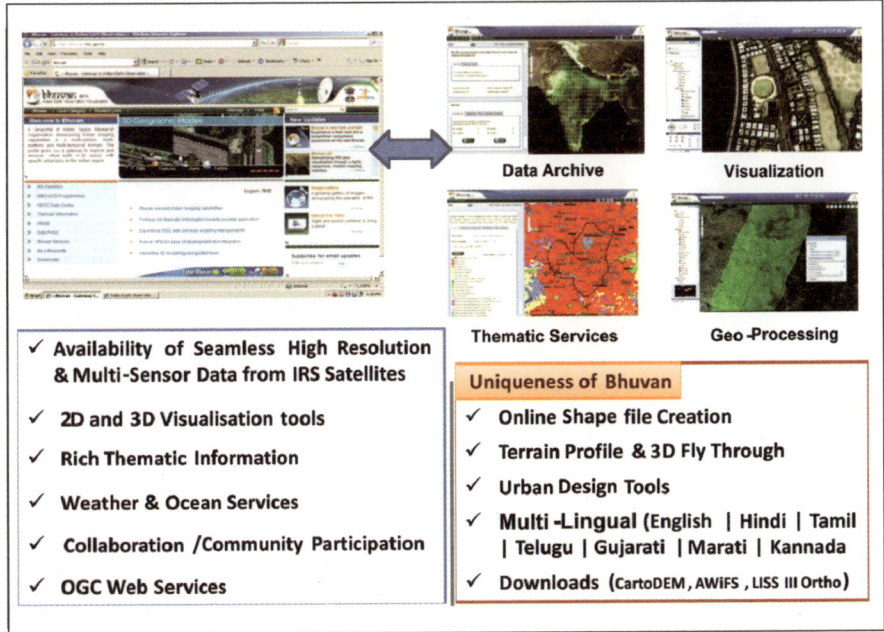

Fig. 3.55: Bhuvan – A Unique Gateway to Indian Earth Observation Data & Services

in the oceans, it is said "time is of essence," as it is often, only by time can we get to a location, especially on the deep seafloor or in the deeper parts of the water column that are out of reach of satellites, global positioning or otherwise. Accurate clocks and accurate timing of the travel of acoustic pulses are critical. In addition to that, Digital Multimedia Broadcasting (DMB) Satellite (S-DMB) using a digital radio transmission technology can provide a solution to many technology challenges when coupled with Remote sensing technologies, Geo-spatial analysis on top of GIS.

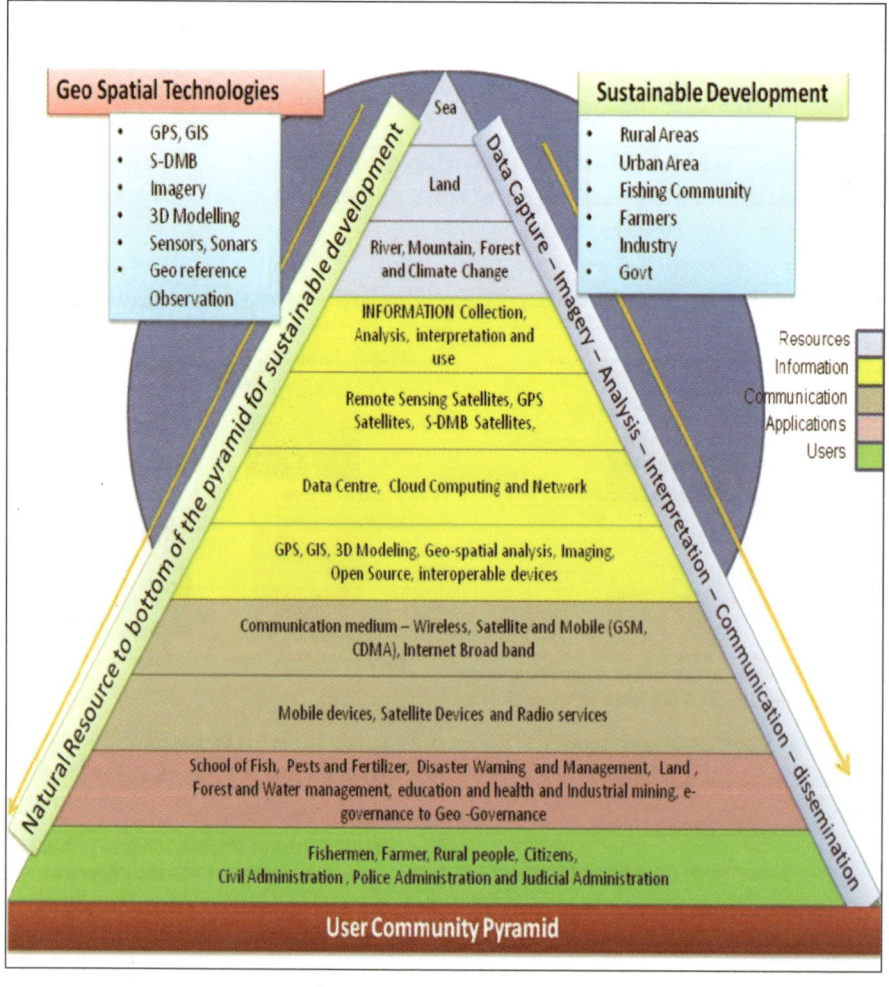

Fig. 3.56: User Community Pyramid

User Community Pyramid – Technology Integration for Sustainable Development

Visualization of Geo-Spatial Pyramid structure linking data acquisition, information, value addition, knowledge and wisdom and finally dissemination to the targeted users is presented in the Figure 3.56. They are called User Community which forms the bottom of the pyramid, and the user is the vital link for all economic activities.

Now, we need to refocus on how we are using technology of the 21st century to solve the problems which are reminiscent of perhaps the 18th or the 19th century. We need to rethink on how the Geo-spatial technologies at our disposal can solve some of the problems of the 3 billion rural population of the world and help them unleash their potential thereby leading to better human life, without damaging the environment around us.

Another challenge we face today is to take urban quality amenities to the 3 billion rural population of the world. This is an urgent challenge which will bridge the divide between the rich and the poor, and urban and rural. The poorest of the world are actually paying the highest per unit cost for basic amenities like clean water, nutritious food and healthcare. How can we overcome this ironic reality of the 21st century? Can the Geospatial Community champion the missions like:

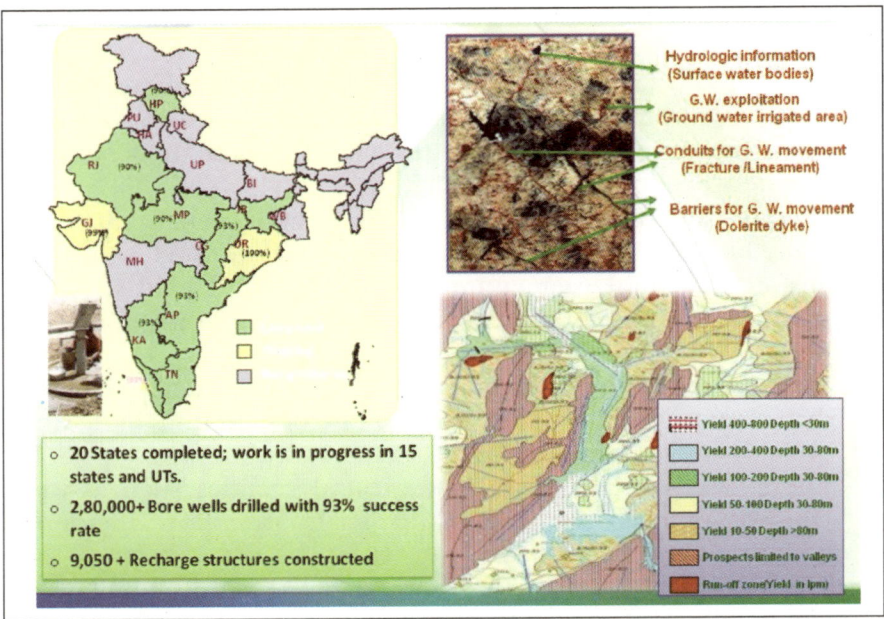

Fig. 3.57: Geospatial Information on Ground Water and Recharge

a) Helping identify the state of potable water availability in the regions. This can include the parameters of both over and underground water supply, pollution status, water borne disease patterns and usage data. This can be the basis for the Potable Water GRID.

b) Helping identify "hot spots" for local energy generation capacity. This can include energy from waste, energy from bio fuels which can be grown in wastelands, small scale hydro plants etc which can empower the local communities. This can be the basis for the Local Energy GRID.

Environment

The third area which is critical in empowering 3 billion people is that of environment. Two factors are noteworthy here, first that the changing environment worst affects the poorest. Nearly, half of the population of the world lives within 150-200 kms of the sea coast and factors like Global Warming are going to worst hit the impoverished in these areas. Secondly, the economically weaker sections are most vulnerable to environment borne diseases. A classic case being that of how lung cancer was predominant in women who were using conventional wood stoves in the rural areas of India.

a) Accurately mapping the carbon balance of different regions of the world. This would include determining the carbon stock, forest and tree cover which offset GHGs, agro generated GHGs emissions etc. This can accurately predict how much net GHG is being generated by each region and nation which will be a significant improvement over the existing estimation methods that we are following.

b) Helping identify optimal locations for renewable energy at a small scale, like wind, solar and cultivation of bio fuels. How can you answer if a rural user community asks you, "I plan to install number of solar panels in my industrial set up in my house and housing complex. Tell me whether it is feasible to install solar panels in the selected locations? What is the frequency of natural disturbances to these solar panels in case of high rain, high winds or high snow conditions? Will it be worthwhile to adopt Solar Energy or Wind based on the ground realities such as wind condition, rain condition and solar radiation incidence conditions and forecasting?

c) Throughout the world, the thermal power plants functioning generate tonnes and tonnes of fly ash. Can you give a fly ash map

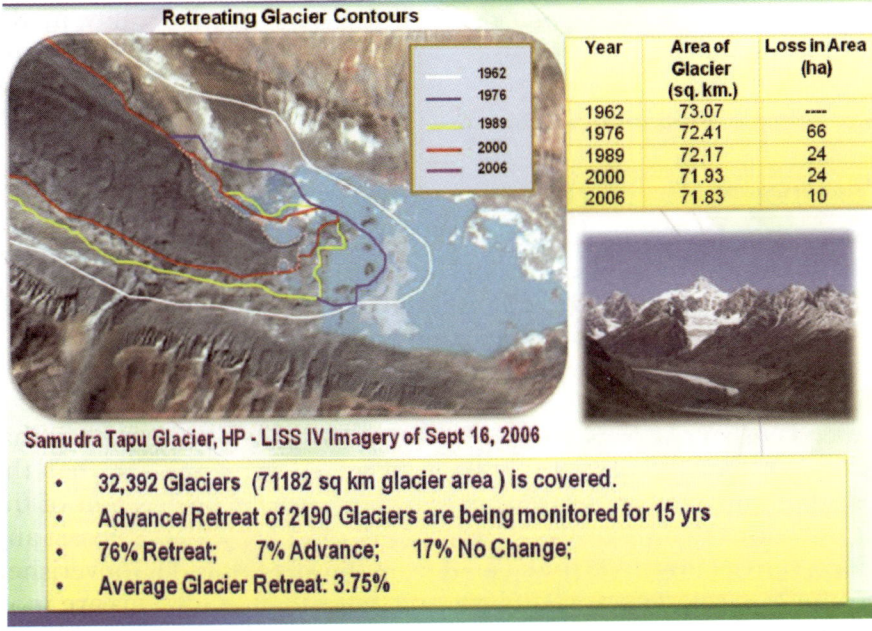

Retreating Glacier Contours

Year	Area of Glacier (sq. km.)	Loss in Area (ha)
1962	73.07	----
1976	72.41	66
1989	72.17	24
2000	71.93	24
2006	71.83	10

Legend:
- 1962
- 1976
- 1989
- 2000
- 2006

Samudra Tapu Glacier, HP - LISS IV Imagery of Sept 16, 2006

- 32,392 Glaciers (71182 sq km glacier area) is covered.
- Advance/Retreat of 2190 Glaciers are being monitored for 15 yrs
- 76% Retreat; 7% Advance; 17% No Change;
- Average Glacier Retreat: 3.75%

Fig. 3.58: Monitoring of Himalayan Glaciers

and municipal waste map? Which will enable me to convert fly ash into building material and fertilizer for effective utilization of these resources?

d) Predicting natural calamities like hurricanes, volcanic eruptions and earthquakes (using the under research surface temperature variations) etc.

e) Can you provide data on a surprise natural wealth, which can become national wealth benefiting thousands? Example, recently detected underwater rivers or precious mineral resources.

f) With more than 17 parameters of geological, magnetic and earth crust movement parameters available, can geospatial community help in predicting an earthquake, atleast one day earlier, ideally seven days before.

g) Can you give every nation a monthly status of deep forest and forest conditions including quantum of oxygen generated?

h) Can you give a map of glaciers of the world, and whether they are in good shape or not?

i) Can you find out which islands are submerging into the sea and to what level merging is taking place due to global warming?

To make all the user community pyramid data useful, irrespective of the nation and the region, it is vital to make it available in commonly acceptable standards and formats which are interoperable protocol standards.

3.8.4 Space Missions of India

Chandrayaan – I

After the Government of India approved its moon mission Chandrayaan-I in November 2003, ISRO successfully launched PSLV-C11 on 22 October 2008. The scientific objective of this mission is high resolution remote sensing in near infra red, low and high energy X-rays region, to conduct chemical and mineralogical mapping of lunar surface, search for water-ice etc. Lunar mission spacecraft carried eleven payloads, five payloads from India and rest from other countries. Satellite placed at 100 km circular orbit around the Moon (Fig 3.59) and instruments gave valuable inputs of moon's surface like distribution of minerals, water craters, etc. On 14 November 2008, the Moon Impact Probe (MIP) ejected from the aircraft and hard landed on the lunar surface near the South Polar Region and placed the Indian Tricolour. The main objective of the MIP is the scientific exploration of the Moon near range, to design, develop and demonstrate technologies required for impacting a probe at the desired location on the Moon and to qualify some of the techniques

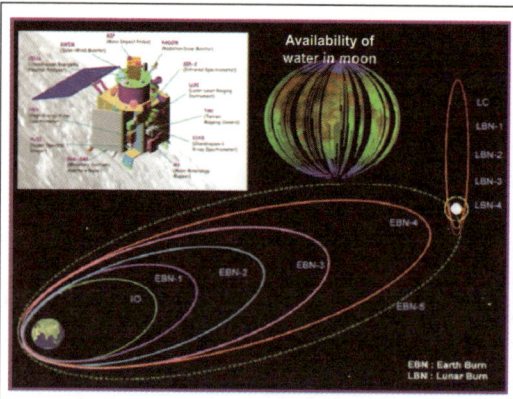

Payloads from ISRO, India

Hyper Spectral Imager (HySI) (0.4-0.9μm)

Terrain Mapping Camera (TMC) (0.5-0.75 μm)

Lunar Laser Ranging Instrument (LLRI)

High energy X-γ ray spectrometer (HEX) (10-200KeV)

Moon Impact Probe

Payloads from Other Countries

Low energy X-ray (LEX) (UK/ESA)

SIR 2 (Max Plank Inst. Germany under MOU with ESA)

Mini SAR (NASA, USA)

Moon Mineralogy Mapper(JPL, NASA)

Sub Kiloelectronvolt Atom Reflecting Analyser (SARA): (Sweden)

Radiation Dose Monitor (RADOM)

Important Milestones of Chandrayaan

Satellite placed at 100 kms circular orbit around the Moon

Instruments gave valuable inputs of moon's surface like - Distribution of minerals, Water Craters, etc.

Fig. 3.59: Chandrayaan – I Mission

required for future soft-landing missions.

Satellite Recovery Mission

The human space flight is the next logical step for India and ISRO is going to put two Indians in an orbit around the Earth in this programme. The flight is likely to take place in 2016-2017. To prove the re-entry and recovery, an experimental satellite called Space Recovery capsule Experiment (SRE) was launched into orbit by PSLV C7. This satellite safely returned to earth twelve days later and splashed on the Bay of Bengal and the module was fully recovered. The experimental details are shown in Figure 3.60. This experiment demonstrated India's capability for re-entry and recovery of space module.

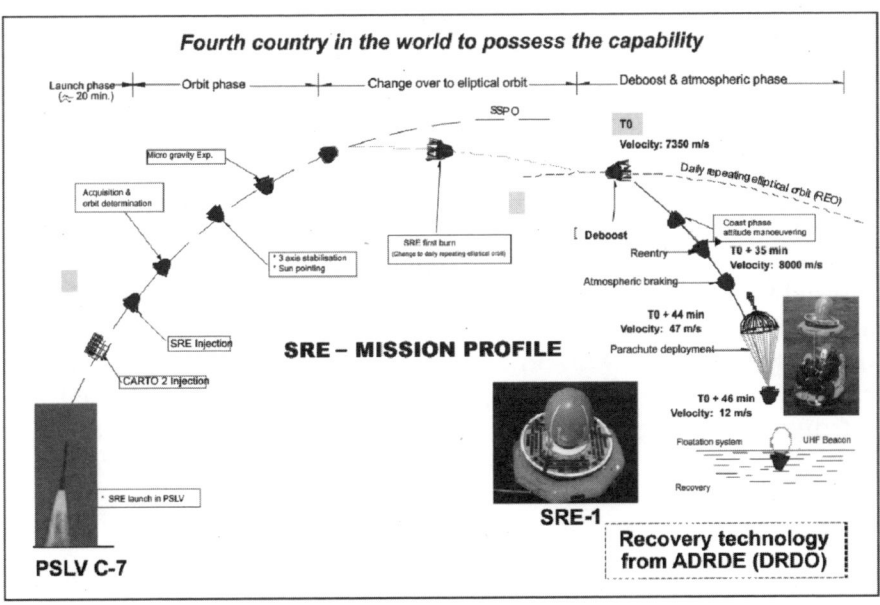

Fig. 3.60: Indian Space Capsule Recovery Mission

Manned Mission

The major objective of manned mission programme is to develop the fully autonomous three-ton ISRO Orbital Vehicle spaceship to carry a two member crew to orbit and safe return to Earth after a mission duration of few orbits to two days. The extendable version of the spaceship will allow flights up to seven days, rendezvous and docking capability with space stations or orbital platform. ISRO plans to use

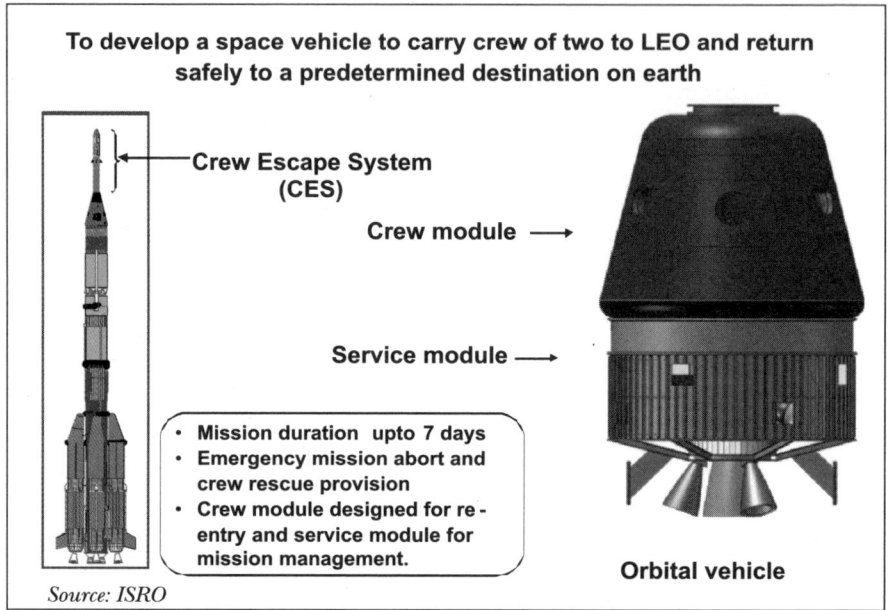

To develop a space vehicle to carry crew of two to LEO and return safely to a predetermined destination on earth

Crew Escape System (CES)

Crew module →

Service module →

• Mission duration upto 7 days
• Emergency mission abort and crew rescue provision
• Crew module designed for re-entry and service module for mission management.

Orbital vehicle

Source: ISRO

Fig. 3.61: India's Proposed Human Space Flight Programme

the GSLV-Mk II launcher (Mark two is Geosynchronous Satellite Launch Vehicle-II launcher with an indigenous cryogenic engine) for Orbital Vehicle (OV) (Figure 3.61). The rocket will inject the OV into an orbit, 300 km - 400 km from the Earth at about 16 minutes after lift-off from the Satish Dhawan Space Centre (SDSC), Sriharikota. The capsule would return for a splashdown in the Bay of Bengal. An astronaut training centre is expected to be set up by ISRO in Bangalore to prepare personnel both for first orbital flights onboard an OV and for future manned Moon missions.

Overview of India's Space Missions and Technologies

ISRO has mastered solid and liquid propulsion technologies through PSLV booster and Vikas engine as well as the upper stages both solid and liquid. Indigenous Cryogenic Upper Stage (CUS) is going through final proving trials for GSLV. ISRO has plans for development of semi cryogenic engine booster and high thrust cryogenic engine in the next five years. GSLV Mk III will provide heavy lift capability for geosynchronous missions. Development of air breathing scramjet engines will be necessary for reusable launch vehicles in the years to come. Both ISRO and DRDO are developing air breathing engines and are planning to demonstrate a hypersonic

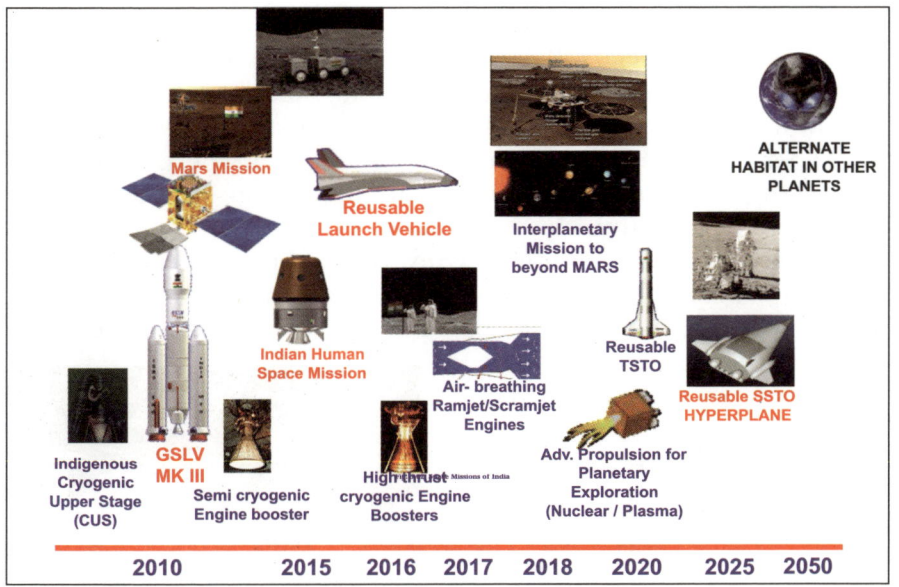

Fig. 3.62 Space Missions of India

flight. As a long term strategy, it is essential to go for cost effective reusable launch vehicles either Single Stage To Orbit (SSTO) or Two Stage To Orbit (TSTO). ISRO has preferred to go through TSTO route for low cost access to space. An overview of the technologies and systems for the next four decade is given in Figure 3.62.

India's Mars Mission is expected to be launched in November 2013 with the aim to collect important scientific information from the Martian surface. A high-end rocket (PSLV-XL) installed with nine instruments will be used to launch the 1.4 tonne Martian spacecraft from Sriharikota. The spacecraft will be placed in an orbit of 500 x 80,000 km around Mars and has a tentative scientific objective for studying the climate, geology, origin, evolution and sustainability of life on the planet. ISRO has planned to establish the entire setup organised by mid-October 2013 so that the launch can take place in November 2013, which has been chosen because the earth and the red planet will be closest to each other.

3.9 SPACE EXPLORATION

3.9.1 Space Based Solar Power

Studies have indicated that the availability of fossil fuels like oil and gas for power generation will be exhausted by 2075 and coal by 2100.

Therefore, to meet the looming energy crisis, the development and large-scale commercial utilization of outer space has been suggested, with the construction of photovoltaic solar power satellites generating electric power for use on earth. Solar energy is available for 99 per cent of the time in an orbit above earth, where 1.43 KW of solar energy illuminates any one square metre considerably greater than that received on earth's surface.

Large solar power stations convert solar flux into microwave energy and beam it down to receiving stations at offshore locations on earth. However, the construction of SPS in space would necessitate the use of Hyperplane, a heavy lift high efficiency space cargo vehicle, using advanced aerospace technologies for revenue-earning mass missions in space. Studies have estimated that one SPS generating about 1000 MW would require 12 sq km array of photovoltaic cells and would weigh 10,000 tonnes (Figure 3.63). Such SPS would take about three years for construction in space using a fleet of Hyperplanes to place construction materials in a low earth orbit.

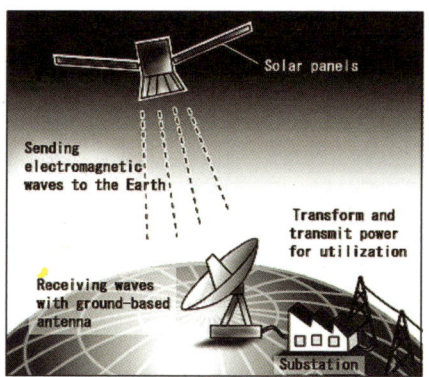

Fig. 3.63: Solar Power Satellite

Kalam-NSS Initiative on Space Solar Power Satellite

During the International Space Development Conference (ISDC) on 04 November 2010 by National Space Society (NSS) of USA, the possibilities of harvesting energy from space through space based solar power were discussed. The experts from the international space community got interested which led to the evolution of the idea of Kalam-NSS Energy Initiative for Space Solar Power Satellite with focus on convergence of competencies from different nations towards the realization of a futuristic mission for green energy from space. This initiative was declared to the global audience by National Space Society.

The sun, as it is known, radiates about 10 trillion times the energy which humans consume across the world today. If we are able to extract even a small portion of this energy from the sun, it would be sufficient to secure the energy demands of our future. Space based solar power has many advantages over traditional terrestrial based solar plants. The level of Solar Irradiance is about 1.4 times in extraterrestrial level than at the surface of the earth. Collection time for the space based solar power plant is 24 hours as compared to the 6-8 hours of surface based solar power plants. These space based solar power plants are not affected by the weather conditions which may bring down the efficiency in case of terrestrial power plant. Thus space based solar power plant would be far more effective in their efficiency and power generation than the land based systems. There are three major focus areas in the space based solar power plant. First component is the space based solar power plant; second is the earth based collection system; and the third important aspect is the medium of transmission from space to earth.

The aspect of safety and efficiency has to be paramount in the way energy is transmitted from space back to earth either through microwave or any other technology like laser technology. Careful research of the impact and safety concerns would have to be conducted. One way to increase safety and improve efficiency could be the evolution of nano-packs, which are reusable, and can move like small batteries carrying charge back and forth from space solar station to ground reception. Another approach could be to make the reception centres as pre-designated offshore sites to reduce safety issues. Another important factor is the cost of the space based power plant, which given the current launch technologies, would be very high and needs to be brought down. Among the largest cost components of installing a space based solar power plant will be the launching cost of the components into the orbit.

Nano energy packs

There is a requirement for the transmission of solar power from space to earth that may be safe and cost-effective. Science has a history of making the impossible today, a reality of tomorrow. If the best brains work together, all these impossible dreamer's ideas will become a reality and the world will become the best place to live. Having established that space is another destination for man's quest for renewable energy, new ideas are to be

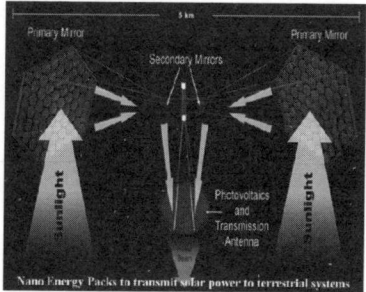

Nano Energy Packs to transmit solar power to terrestrial systems

implemented for transmission of energy from space. Though the approach of transmission of energy from space via micro waves and laser beams, new concepts like "nano energy packs", which are like nano batteries, back and forth between space and earth, could be attempted to transport the energy. These nano energy packs may contain materials hitherto unknown but would store the energy through reversible chemical reaction or perhaps electrical reaction and when brought back to earth, can deliver energy per kg of payload touching several hundreds of watt hours.

3.9.2 Repair and Refuelling of Satellites in Orbit

There are hundreds of operational commercial satellites and government spacecraft currently in orbit, many of which will run out of fuel long before they sustain electronics or other systems failures. Enormous amount has been spent on sending the satellites to orbit. The goal is to find an optimum solution for refuelling, repairing and servicing spacecraft in orbit. The advancements in Robotics and Artificial Intelligence will give way to performing such tasks in space. The humanoid robots could be deployed in space for undertaking any repair of the satellites and for extending the lifetime of the satellites.

3.9.3 Collecting Space Debris

Space debris also known as orbital debris, space junk, and space waste is the collection of objects in orbit around Earth that were created by humans but no longer serve any useful purpose. These objects consist of everything from spent rocket stages and defunct satellites to explosion and collision fragments. The debris includes dust from solid rocket motors, surface degradation products such as paint flakes, coolant released by RORSAT nuclear powered satellites, clusters of small needles, and objects released due to the impact of micrometeoroids or fairly small debris onto spacecraft. As the orbits of these objects often overlap the trajectories of spacecraft, debris is a potential collision risk.

The vast majority of the estimated tens of millions of pieces of space debris are small particles, like paint flakes and solid rocket fuel slag. Impacts of these particles causes erosive damage similar to sandblasting. The majority of this damage can be mitigated through the use of a technique originally developed to protect spacecraft from micrometeorites by adding a thin layer of metal foil outside the main spacecraft body. Impacts take place at such high velocities that the debris is vaporized when it collides with the foil, and the resulting plasma spreads out quickly enough so that it does not cause serious damage to the inner wall. However, not all parts of a spacecraft may be protected in this manner, e.g. solar panels and optical devices.

There are several ways of dealing with debris. In order to mitigate the generation of additional space debris, a number of measures have been proposed. The passivation of spent upper stages by the release of residual fuels is aimed at reducing the risk of on-orbit explosions that could generate thousands of additional debris objects. Another process is self removal. Geostationary satellites will be able to remove themselves to a "graveyard orbit" at the end of their lives. It has been demonstrated that the selected orbital areas do not sufficiently protect GEO lanes from debris, although a response has not yet been formulated. Rocket boosters and some satellites retain enough fuel to allow them to power themselves into a decaying orbit. In cases a direct (and controlled) de-orbit would require too much fuel, a satellite can also be brought to an orbit where atmospheric drag would cause it to de-orbit after some years. Another proposed solution is to attach an electrodynamic tether to the spacecraft on launch. At the end of its lifetime, it is rolled out and slows down the spacecraft. It has also been proposed that booster stages include a sail-like attachment to the same end. The vast majority of space debris, especially smaller debris, cannot be removed under its own power. A variety of proposals have been made to directly remove such material from orbit. These range from large spacecraft capture and hazard mitigation to "laser brooms" for removing small pieces of debris.

3.9.4 Space Colony

With the world population topping 4 billion people and expected to reach over 30 bn within the next 100 years, efforts are being made to explore the possibility of accommodating people and building a city in space which could house several thousand inhabitants and boast of an environment identical to Earth. The colony will have air, water, lakes and mountains. It will have gravity and similar atmospheric structure, air pressure and temperature as existing on earth and people will be able to stand

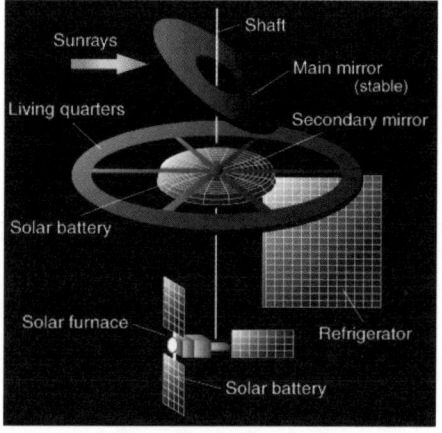

Fig. 3.64: Space Colony

and walk exactly as they do on earth. Light would be provided by massive reflective mirrors affixed on satellites that reflect sunrays to the colony. Living quarters would be a donut shaped area. Colony would also contain areas for agriculture, animals and plants and would weigh about 10 million tonnes (Figure 3.64).

Recently, Space Settlement contest was organised by NASA wherein 355 designs were submitted from fourteen countries. The Grand Prize for the 2011 NASA/NSS Space Settlement Contest went to a team of seven high school students from Punjab, India, for their double-torus space settlement design called Hyperion. The Hyperion Space Settlement has a diameter of 1.8 km and would provide a safe and pleasant living and working environment for 18,000 full time residents and an additional population (not to exceed 2,000) of business and official visitors, guests of residents, and vacationers. The settlement would be constructed primarily from lunar materials and be located at the Earth-Moon L4 libration point. The French Mathematician LaGrange enunciated libration points at five locations where the gravity forces between earth and moon become null. L1, L2 and L3 are in the same line of earth and moon axis and the L4 and L5 are equilateral as shown in Figure 3.65.

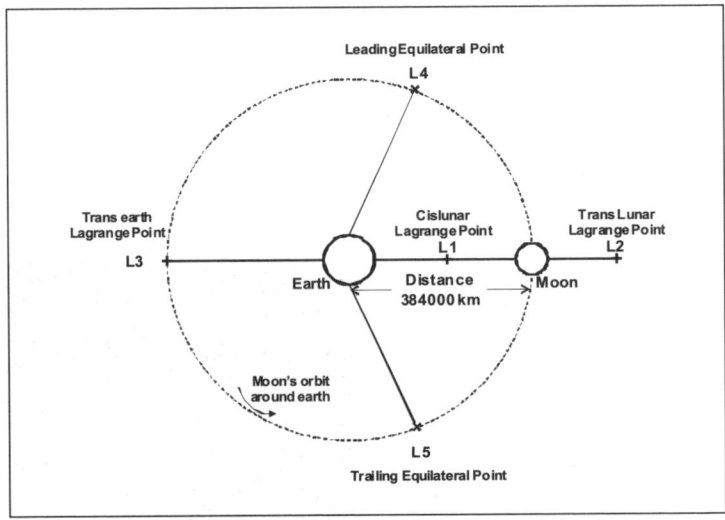

Fig. 3.65: Earth-Moon Lagrange Points

Prof. Gerard K. O'Neil of Princeton University had visualized a space colony in the earth-moon L4 and L5 liberation points and established a unique laboratory to experiment the theory of space colonization. A series of Heavy Lift Launch Vehicles (HLLV's) or

Hyperplanes will carry hundreds of tons of cargo consisting of equipment to Low Earth Orbit (LEO). The assembly and transportation of this equipment meant for a space colony would be done by a ferry service called Space Tug. The new idea is to have a launch base at the moon which has less gravity as compared to the earth (1/6th). This will reduce both the weight of the launch vehicle as well as time for the flight.

There are two possibilities; one directly taking payloads to the space colony from LEO (Figure 3.66) and the other route is to establish a minimum facility on the moon for material mining and reaching the L4/L5 point through catch point L2. This is because if the space colony is to be built at L4/L5 point between the earth and moon, it is thought better to transport the necessary building materials from the moon, due to its less gravity and because rockets taking off from there would require less fuel. Because of this factor, a factory would need to be set up on the moon before work on the space colony could begin. Materials from the moon can be carried to the colony through lunar materials transfer vehicle and construction in space. The Milky Way now awaits for our conquest. It will happen in the next century that people will be living in space colonies and will be travelling to earth for holidays.

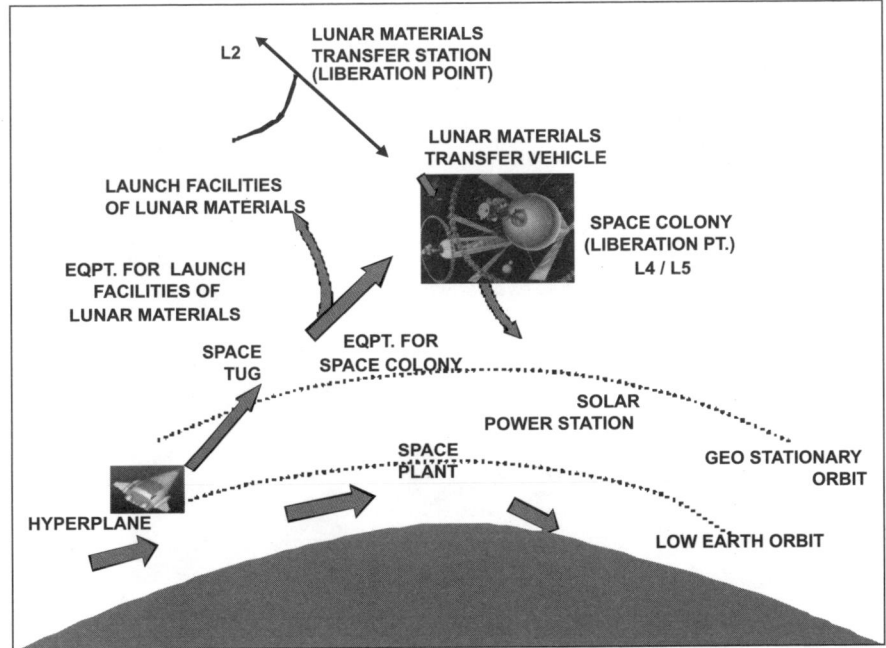

Fig. 3.66: Space Colony at Liberation Point

3.9.5 Space Industry

Space offers an enterprise for the future generation with next industrial revolution (Figure 3.67). Availability of exotic resources and low gravity manufacturing on moon and Mars have tremendous prospects for mankind. Mining in planets would need innovative methods for exploring, processing and transporting large quantities of rare materials to earth. The moon could become a potential transportation hub for interplanetary travel. The moon's sky is clear to waves of all frequencies. With interplanetary communication systems located on the far side, the moon would also shield these communication stations from the continuous radio emissions from the earth. Hence, the moon has the potential to become a launch base for interplanetary travel and a 'Telecommunications Hub'.

Man's quest for perennial sources of clean energy such as solar and other renewable energies and thermonuclear fusion would be filled through mining/exploration on the moon. Large deposits of Helium-3 on the moon and Mars provide a solution for future energy demand. Also, the dry ice deposits on the planets would be a source of fuel for rocket engines. 100 kg of Helium-3 would have a coal equivalent value of $140 million. Access to lunar Helium-3 at competitive cost potentially offers an environmentally benign means of helping meet an anticipated ninefold or higher increase in energy demand by 2050. Samples collected in 1969 by Neil Armstrong during the first lunar landing showed that Helium-3 concentrations in lunar

- MANUFACTURING IN REDUCED GRAVITY

- MINING OF SCARCE RESOURCES

- LARGE RESOURCE OF HELIUM -3 IN MOON & MARS FOR POWER GENERATION

- DRY ICE DEPOSITS IN MOON AND MARS AS SOURCE OF FUEL ROCKET ENGINE

- MOON TO BE USED AS A HUB OF TRANSPORTATION TO LIBRATION POINTS & OTHER PLANETS

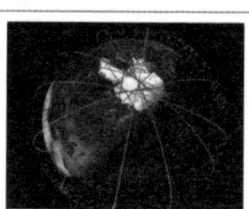
Celestial Broom

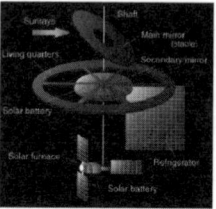
Space Colony

Mining in Planets

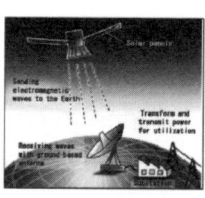
Solar Power Satellite

Fig. 3.67: The Future Revolution

soil are at least 13 parts per billion (ppb) by weight. Levels may range from 20 to 30 ppb in undisturbed soils. But at a projected value of $1410 per gm, 100 kg of Helium-3 would be worth about $141 million. The highest concentrations are in the lunar maria; about half the He3 is deposited in 20 per cent of the lunar surface covered by the maria. It is believed that there are large deposits of He3 that have been deposited by solar wind in the lunar soil. Since the lunar soil has been stirred by collisions with meteorites, possible availability of He3 on moon would be down to depths of several metres. That 1 million metric tonnes of He3, when reacted with  deuterium, would generate about 20,000 terrawatt-years (one trillion (10 to 12th power) watt-years) of thermal energy. Hence, Helium 3 is the future (Figure 3.68).

3.9.6 World Space Vision 2050

Challenges for Space Community

While the entire world is actively involved in futuristic research, development and mission efforts related to exploration of solar system and beyond, search for extra terrestrial life, further moon and mars exploration, the following are some of the challenges that would be faced by the space community:

1. The problem facing human life on earth and protection and sustainability of environment, requires many areas of research like integrated study of atmosphere, accurate forecasting and predictions of climate and weather, breakthroughs in earthquake forecasting, making energy independence and finding a solution to water problems for the growing population of the world and fast track improvement of education and healthcare system throughout the globe. They require integrated global approaches and synergy of all space faring nations.
2. While in the last fifty years, tremendous progress has been made by space faring nations in space science, technology and applications, our understanding of our own planet is limited, and we have much less knowledge of moon, mars, other planets and the solar system and beyond, planning and conducting such missions need lead times of the order of 10-15 years and individual nations find it difficult to increase the rate of

Fig. 3.68: Helium 3 – The Future

exploration momentum with the available technology level, funds and human resources.

3. In the last fifty years, space-faring nations have developed core competencies like strong application orientation in India, planetary mission and space station capabilities in USA and Russia, a model of nations working together in Europe and manned space mission capability in China. The launch vehicle and spacecraft expertise exist in many of the space faring countries. There have been a number of missions with bilateral and multilateral partnership with several complimentary efforts. The question is whether the international cooperation is consistent with the challenges of the next fifty years. Can we graduate in the ensuing years to partnership missions among space faring nations for the benefit of entire humanity using the core competence of multiple nations, financial sharing and management mechanism with overall goals like solar power satellite, industrial belt on the Moon and the habitat on the Mars? Political will and concern for the well being of planet earth have to be the focus for such international missions.

4. Missions can be evolved but one important constraint among space faring nations is the non-availability of the right type of young human resources. What we need is a global vision that would attract the imagination of the young and inspire them to achieve their own dreams . Spotting young talent and training them requires attention internationally.

5. There was a strong feeling among many participants that enough communication is not ensured by the performing space community with general public on the system they have created using space technology for the welfare of humanity.

Paradigm Shift in Foreseeing Space of Next Fifty Years

The present capabilities of major space faring nations are not optimally utilized. The launch vehicles of the world, the spacecraft of the world, the application potential of the world, the space scientific research potential of the world and above all the huge costs envisaged for space 2050 programmes would call for a certain "paradigm shift" in nations to work together to bring the benefits of space to humanity as a whole. This is possible only if we have a mutual cooperation of each nation contributing substantially to technology and resource. Examples of such cooperation are the Joint Venture programme between India and Russia with the shared funding of $300 million that has resulted in the development leading to the production of the world's first supersonic cruise missile called BrahMos in the defence sector and the Pan African e-Network initiative costing over $100 million, for connecting 53 Pan African Nations for providing education, healthcare and e-governance services.

Similarly USA, Europe and other countries have many experiences in mutual cooperation. These experiences give the confidence that international cooperation indeed can accelerate the application of space science and technology leading to fast results for societal application. Such international cooperation aids the security dimension in space. Such an accomplishment of a goal would enable taking up mass missions that were hitherto not in the realm of individual nations. A world space vision can definitely trigger many young people towards hitherto "impossible" challenges.

Since the dawn of space era in 1957, space science and technology has enhanced man's knowledge of earth, atmosphere and outer space. It has improved the quality of life of the human race. With sound background and technological progress in space systems made by the world countries, World Space community can evolve the space vision for the next fifty years, consolidating these benefits and expand them further to address crucial issues faced by humanity in energy, environment, water and minerals. The World Space Vision 2050 (Figure 3.69) would enhance the quality of human life, inspire the spirit of space exploration, expand the horizons of knowledge, and ensure space security for all the nations of the world. The vision will have the following three components:

1. Large Scale Societal missions and Low cost access to space: There is definitely a need for space-faring nations to work together in order to develop reusable launch vehicles which can bring down the cost of payload in orbit from the present US $ 20,000 per kg to US $ 2000 per kg and eventually to $200 per kg.

2. Comprehensive space security: Creation of an International Space Force (ISF) made up of all space faring nations wishing to participate and contribute to protect world space assets in a manner which will enable peaceful use of space on a global cooperative basis.

3. Space exploration and current application missions: Space exploration mission for material like Helium3, water and life and the current application missions in telecommunication, remote sensing and other societal applications and considering Earth-moon-mars as a single economic complex for the benefit of humanity.

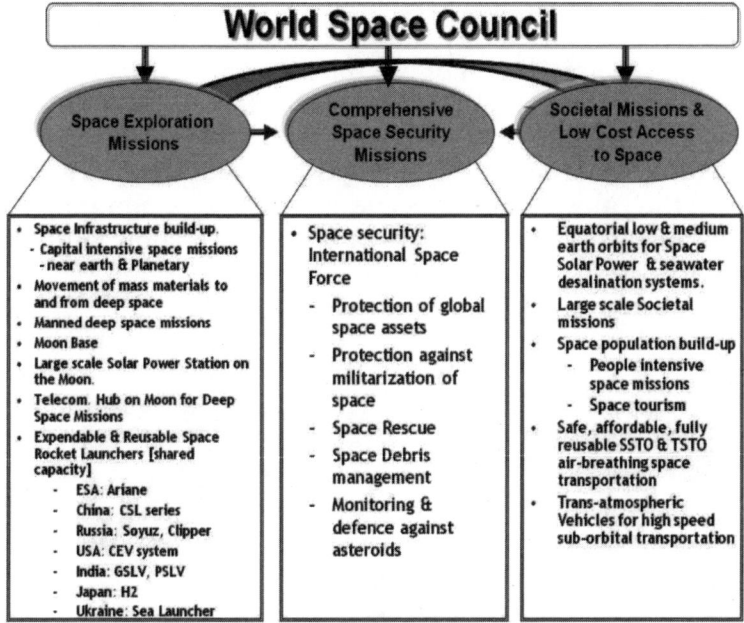

Fig. 3.69: World Space Vision 2050

International space community can come together on a common platform to evolve and implement World Space Council, consisting of space faring nations, to formulate and implement World Space Vision 2050. The World Space Council with global participation could oversee the planning and implementation of exploration, energy and societal missions. Such a unified approach will enable the world to see a quantum jump in the progress in space science and technology for the benefit of all the nations of the world. The need of the hour is a step function as a global space initiative to implement a World Space Vision and missions for an enhanced quality of life for a peaceful and safe world.

3.9.7 Future Space Frontiers

"The human race will get out of its cradle – the earth to explore the new frontiers of space, like a child gets out of its cradle to explore things around it."

Beyond 2020, one can imagine that space tourism will become popular, which will include visits to the moon and further to Mars using reusable vehicles such as Hyperplane offering cost-effective transportation. Space factories on the moon and mars will be built for mining rare resources such as Helium-3 for generating energy. Moon will also become the future launch station for exploring the space frontiers as the gravity of moon is one sixth of the earth making the cost lesser. With the increasing demand for electric power, many nations will depend on Solar Power Satellite (SPS) which could be built in space and launched in a geo-stationary orbit. The SPS will become the most cost effective means of generating power. Establishment of a space colony at L4/L5 liberation points will be realized from the base station on the moon. New findings of earth-like habitable planets will further enhance the establishment of alternate habitat on the other planets such as Kepler 22B and GJ 667C.

Fig. 3.70: Space Frontiers

The future space travel can be interplanetary between planet to planet, interstellar between star to star, intergalactic between galaxy to galaxy and travel faster than light. The transport network for the interplanetary travel within our solar system could be a pathway gravitationally determined such that energy consumption is very little. While travelling one could think of generating electricity to propel matter with little thrust operating for long time.

It could even be solar sails using Sun's radiation or laser, generating power from nuclear thermal or solar thermal engines. For interstellar travel, fusion rockets, anti-matter rockets, laser propulsion, and radiation propulsion can be used. Long trips of interstellar travel could extend the life span of human beings. A bubble of space-time, if it can be made to travel faster than light, we could place a spaceship inside that bubble. 200 scientists from 13 countries are working at OPERA (Oscillation Project with Emulsion-tRacking Apparatus) to find out whether Neutrinos can travel at a velocity greater than the speed of light in vacuum.

A day will come when the Human Being will settle in a space colony and in other planets beyond the solar system, and ultimately conquer the Milky Way. The human race will come back to the beautiful Mother Earth to spend holidays.

3.10 MISSILE TECHNOLOGY

The use of missiles dates back to the Vedic age in India. Indian warriors of that period had used "Astras" as missiles in various forms in the subcontinent. Proofs of which can be drawn from a number of epics such as the Mahabharata and the Ramayana. The deadly weapons and the scriptures regarding the technology were later concealed to prevent any future happenings of the "Idikasa" (an important historic part in the history of Indian religion) for the sake of the survival of mankind. In olden days, the "Astra" (a supernatural weapon) was controlled by "Mantras" (a sound, syllable, or group of words that are considered capable of creating transformation) that could be correlated to the mission control software of the modern day missiles.

The word missile comes from the Latin verb mittere, meaning "to send". In common military parlance, the word missile describes a powered, guided munition, whilst the word "rocket" describes a powered, unguided munition. Unpowered, guided munitions are known as guided bombs. A common further sub-division is to consider ballistic missile to mean a munition that follows a ballistic trajectory and cruise missile to describe a munition that generates aerodynamic lift with continuous propulsion throughout the flight.

Guided Missile

War weaponry had gone through various phases of development, particularly during the twentieth century with new weapons and platforms. Most famous of these were the V1 and V2, both of which used a simple mechanical autopilot to keep the missile flying along a pre-chosen route. Lesser known was a series of anti-ship and anti-aircraft missiles, typically based on a simple radio control system directed by the operator. However, these early systems had a high failure rate so they were unreliable. The most significant developments were the V2 rocket, the first guided missile developed by Von Braun, Germany in the Second World War, and Tomahawk cruise missiles of the USA used in the Gulf war. Today, BrahMos, a supersonic cruise missile which is versatile and capable of multi-role, multi-target and multiple missions has emerged. From the world's first rocket in 1792, India travelled a long way till 2001 with BRAHMOS, the world leader of the cruise missile family, and 2012 with Agni V, a 5000 km range strategic missile. From human warfare, we have come to intelligent and autonomous systems with faster operational capabilities to utilise deep sea, space and networked sensors and weapons.

3.10.1 India's Missile Programme

Defence Research in India had a modest beginning in 1958. Dr. DS Kothari, a well known educationist, a renowned physicist and a highly successful leader and organiser took charge as the Head and Scientific Advisor to Raksha Mantri and steered the research. Over the years, DRDO has seen multi-directional growth and has proven its competence to produce state-of-the-art systems and technologies in diverse disciplines such as missile, armaments, combat vehicles, engineering, electronic warfare Radar, aeronautics, materials etc.

The mission of Defence Research and Development Organisation (DRDO) is to design, develop and lead to production of state of the art equipment and weapon systems for the Armed Forces and to achieve self-reliance in critical technologies. Over the last fifty years, DRDO has established itself as a unique organisation with high quality technologists and scientists in different fields ranging from missiles, electronic warfare, underwater systems to materials including advanced composites, processors, supercomputers, robotics, radars and NBC warfare.

DRDO's Defence Research and Development Laboratory (DRDL) at Hyderabad is the nervecentre for design, development and the lead to production of guided missiles for India. In the 1960s, missile technology started with the development of anti tank and surface to air missiles, initially taking clue from the already deployed Soviet missiles. The major task was to build human resource that can design missiles and develop technologies which are critical to realise the systems. Primary focus was on developing propulsion, structures, control and guidance technologies and systems as well as trajectory design and system engineering. Infrastructure build-up was also necessary to meet this development effort. Though initial development gave some success in the 70s, none of the systems went for production due to the availability of better systems through import. We realised that unless the missiles developed in India are contemporary in terms of technology and performance, they will not be appreciated and accepted by the Indian Armed Forces. A missile technology committee pondered over the situation and evolved the Integrated Guided Missile Development Programme (IGMDP), briefly discussed under Mission Mode Programmes in Part 2.

IGMDP started in 1983 constituting five projects—Surface to Surface Missile–Prithvi; Intermediate Range Ballistic Missile – Agni; Surface to Air Missiles–Trishul and Akash; and Third Generation Anti Tank Guided Missile–Nag. These were state-of-the-art technology missiles to be developed from scratch. India was more than twenty years behind in missile technology when the programme was started. The design of these missiles incorporated high technologies in order to make the missiles contemporary in performance at the time of their induction, compared to world status. It meant the technologies to be chosen had to be developed at a faster pace than the industrially developed nations. The national strengths had to be pooled from academic institutions, R&D organisations and industries. The development of these technologies had to be undertaken totally with our own effort because of the Missile Technology Control Regime (MTCR) and technology denials by developed countries. This challenge was taken up by the whole nation with multi-institutional partnership. There was an urge to combat the control regimes and to establish that "We can do it".

Many new institutions emerged and new management methods were evolved. The result of this indomitable spirit is that today India possesses high technology skills and capabilities in design and development of any type of missile system. It possesses technologies such as supercomputing, computational fluid dynamics, re-entry structures with carbon composites, phase shifters required for multifunction radars for multiple missiles tracking and guidance, high accuracy sensors with embedded software and many others. Self-reliance in critical missile technologies placed our country close to the developed nations.

As a result of these efforts, at present, Agni and Prithvi strategic missiles are operational. Nag has gone through very successful flight trials with its advanced imaging infra-red guidance. Akash had gone through several flight trials and is being inducted by the Indian Air Force as air defence weapon. Trishul, as a Technology Demonstrator project, has established several critical technologies for future development of many systems/subsystems.

Prithvi is a short-range surface-to-surface missile having high accuracy and manoeuvrable trajectory. Manoeuvrable trajectory of the Prithvi ballistic missile made this a unique missile. It can be launched from both mobile launchers and from ships. The trajectories of Prithvi can be manoeuvred in pitch plane as well as yaw plane to make it more deceptive while attacking a target. The ship launched a variant of Prithvi that is known as Dhanush and it also had all the accuracy and lethality of Prithvi. Besides, it had the advantage of being launched from a ship against any land target. Prithvi is powered by a liquid propulsion engine (Figure 3.71).

The role of Prithvi and Dhanush was of tactical/strategic missiles. The guidance and control system of Prithvi was designed and developed by collaborative efforts of R&D institutions, academia and industries. It has an onboard computer with an inertial navigation system and embedded software. Inertial guidance uses sensitive measurement devices to calculate the location of the missile due to

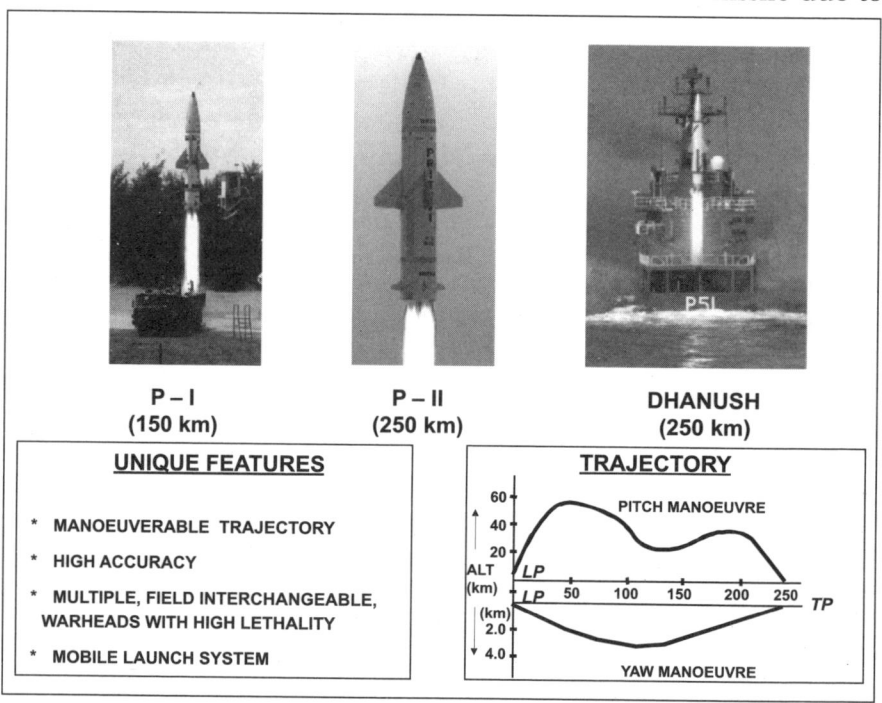

Fig. 3.71: Prithvi Surface to Surface Missile System

the acceleration put on it after leaving a known position. INS systems use solid-state ring laser gyros that are accurate within metres over ranges of 10,000 km.

Agni missile (Sanskrit name for Fire, one of the five elements of nature) is an Intermediate Range Ballistic Missile (IRBM) developed under IGMDP. It was first tested as Technology Demonstrator at the Integrated Test Range in Chandipur in 1989. The variants of Agni namely single stage Agni I for 700 km range and two stage Agni II with 2000 km range have already been inducted in the strategic command. These variants can be deployed from rail system and by road mobile. Agni III is a two stage larger diameter system for a range of 3500 km and Agni IV is an advanced version derived from Agni II using composite casings and newer guidance system for a range of 4000 km. Recently DRDO has tested Agni V for a range of 5000 km by adding a third stage to Agni III (Figure 3.72).

Agni uses solid propellant rocket motors, a number of composite material products such as casing, flux nozzle, re-entry structure. Due to its long range, the missile has to attain an altitude of more than 500 km and returns in the ballistic trajectory at a high speed to make a re-entry in theatmosphere at Mach 14, experiencing a friction temperature of 3000° C. Development of re-entry structure with composite materials, designing the mission parameters and loads at re-entry, test of the re-entry on ground, control and guidance after re-entry posed greater challenges to the missile technologists. Missile trajectory including the re-entry phase is shown in Figure 3.73. In addition, India mastered the parallel processing supercomputer PACE++, computational fluid dynamics with new codes, solid propellant technology, carbon-carbon and composite structures, mission design for long range weapons with number of software packages, integrated data acquisition and management during flight tests, and realisation of wheeler island range with a host of radars, telemetry stations, electro optical systems and network of computing centres, and all these systems and technologies have become an asset for long range missions.

Nag (Sanskrit name for Cobra) is India's third-generation 'Fire-and-Forget' anti-tank guided missile. It is an all-weather, top attack missile with a range of 4 km. The missile uses an 8 kg tandem High Energy Anti Tank (HEAT) warhead capable of defeating modern armour including ERA (Explosive Reactive Armour) and composite armour. Nag uses Imaging Infra-Red (IIR) guidance with day and night capability. The mode of launch for the IIR seeker is LOBL (Lock On Before Launch).

Fig. 3.72: Ballistic Missile Agni

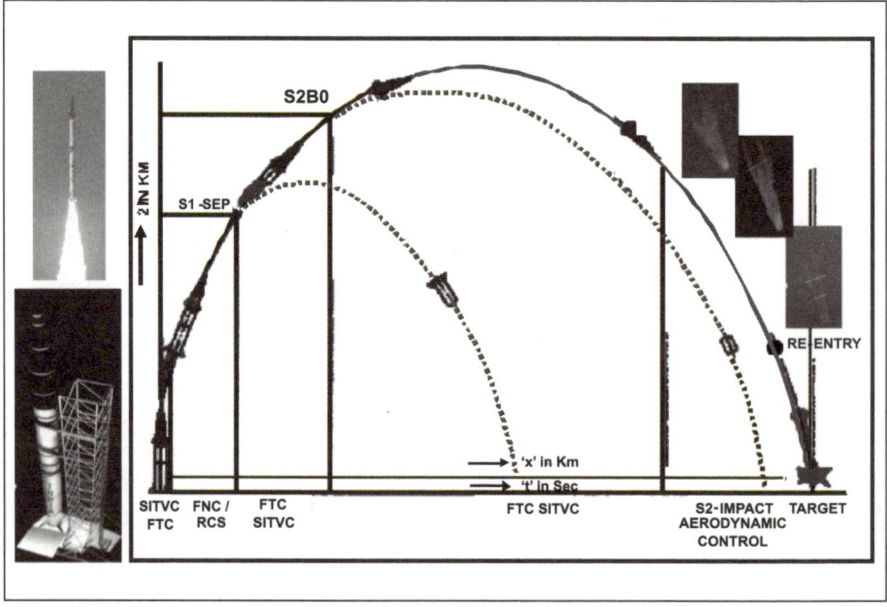

Fig. 3.73: Agni - India's First Long Range Re-entry Vehicle

Nag can be mounted on an infantry vehicle; a helicopter launched version will also be available with integration work being carried out with the HAL Dhruv.

Two versions of Nag are being developed – one for tank to tank warfare and the other one from helicopter to the tank. In the first version, the missiles are carried by specialist carrier vehicles NAMICA (Nag Missile Carrier) equipped with a state-of-the-art thermal imager for target acquisition, a top attack trajectory and a fire and forget system using its Imaging Infrared seeker. NAMICA is a modified BMP-II Infantry Combat Vehicle. The carriers are capable of carrying four ready-to-fire missiles in the observation/launch platform which can be elevated with more missiles available for reloading within the carrier. In the second version, eight missiles are carried by the Advanced Helicopter Dhruv. A nose-mounted thermal imaging system has been developed for guiding the missile's trajectory.

Trishul (Sanskrit name for Trident) is a short range, quick reaction, all-weather surface-to-air missile designed to counter a low-level attack. It has been flight tested in a sea-skimming role and also against moving targets. It has a range of 9 km and is fitted with a 5.5 kg HE-fragmented warhead. Its detection time of target from missile launch is around 6 seconds. The missile can engage targets like aircraft and helicopters, flying between 300 m/s and 500 m/s by using its radar command-to-line-of-sight guidance. It operates in the Ka-

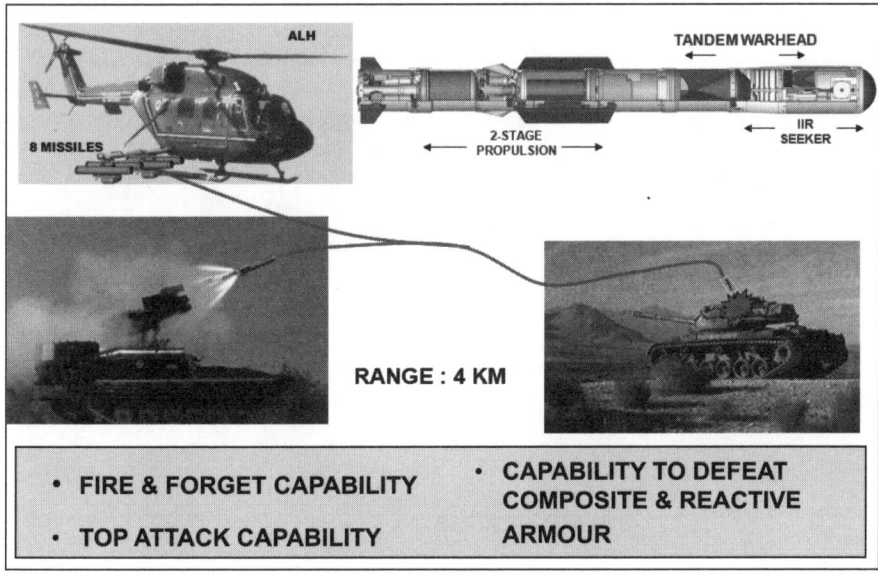

Fig. 3.74: Nag - Third Generation Anti-tank Guided Missile

band (35 GHz), which makes it difficult to jam. In the Ka-band three-beam system, the missile is initially injected into a wide beam, which then hands it over to a medium beam, which passes over to a narrow beam, guiding it to the target. The project has been closed as a technology demonstrator.

Akash (Sanskrit name for Sky) is a medium-range, theatre defence, surface-to-air missile. It has got a multi-target handling capability and Electronics Counter Counter Measures (ECCM) features. It has got a maximum range of 25 km and can fly at an altitude of 30 m (tree top) to 18 km. The missile is capable of multi-target engagement and with integrated identification of friend or foe (IFF). This is a complete mobile system with C4I Network. Akash missile system is guided by phased array radar which can track multiple targets using phase shifters and guide multiple missiles towards these targets simultaneously. Akash uses the Integral Rocket Ramjet propulsion to give a low-volume, low-weight (700 kg launch weight) missile configuration, and has a low reaction time of 15 seconds from detection to missile launch. Akash has gone through several successful flight trials and has been accepted by the Indian Air Force for induction.

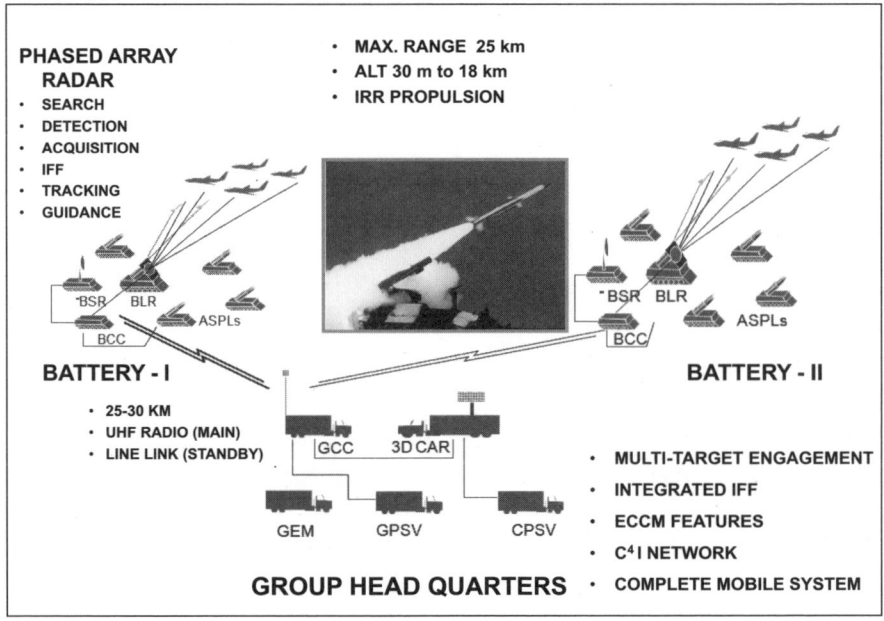

Fig. 3.75: Akash Weapon System

3.10.2 Development of Critical Technologies

Building critical technologies is full of challenges. They are critical for the realisation of important sub systems/systems. Similarly, their management from concept and design to realisation and quality is quite a complex task. The need to develop a critical technology arises from the development of high technology oriented systems. The correct assessment of need, critical technology, plays a decisive role. Based on this assessment, requisites like manpower, infrastructure and capital are made available to realise the set objectives. There could be many sub-objectives in a single critical technology. Therefore, the management of it becomes a challenging task.

Building Capabilities through Knowledge

The development of any new technology is a challenging task. There are many complexities involved in the course of new technology development. There are many influencing factors such as design and development; manufacturing capabilities; quality control and customer satisfaction for the successful development of new technology. Building the requisite capabilities, to deliver the desired technology of product, requires knowledge.

Knowledge is a very versatile tool and its contribution to product value is enormous. The major role of knowledge is attributed to the design which has a major stake in the value addition of any product. All other parameters such as cost, quality, reliability and produceability depend upon design. The major boost on quality improvement revolves around the efficacy of design. The cost effectiveness of the product, its quality, and serviceability are the main causes for generation of demand. The prime role of a design is to satisfy the needs of the end user or customer in the most economical way. If we take any product lifecycle, it can be seen that among the different functions of specification/requirement, design, process, manufacturing, services and product improvement, wealth generation depends mostly on the design phase accounting for 60 per cent of the product life cycle efforts. Hence, indigenous design, i.e., R&D plays a vital role in product realisation.

Realisation of New Product by R&D

Realisation of a new product involves many skills and factors. The most important of all is the proper understanding of the requirement of the end user or customer. To meet a specific requirement, the customer may demand the development of new technologies; thereby technology complexity comes into the picture. There are many other influencing factors to realise new products such as development strategy, risk and uncertainty, quality standards needed, product support and life cycle cost. A good balance among all these factors will see the successful development of a new product or technology. Finally, the product has to be competitive to create its market and for this to happen, all the influencing factors need to be well analysed, understood and addressed. The product should be cost-effective by way of both product cost and life cycle cost.

Challenges in R&D Projects

Technology finds its use in the form of products and the realisation of products is dependent on how we tackle the complexities in technologies. The important factors to be addressed in the complexities in technologies are the new requirements; scarcity of knowledge workers and speciality equipment, resources or priority clashes; concurrent engineering; partnerships; matrix and network organisations; safety and ergonomics; product lifecycle and cost control. There are always some uncertainties in new technology development. It is extremely difficult to estimate the magnitude of task, technology and skills requirement, resource requirement and timeframe. While development is in progress, technology advances very fast and customer preferences undergo greater changes. High technology development involves knowledge workers, specialised equipment, matrix and networked organisations with multi-institutional partnership and concurrent engineering practices. The correct measure of the skills of human resource, cost and schedules is a complex task.

There are many obstacles encountered while progressing a project. Sometimes, the requirement of the user also changes with time and keeping pace with it is the paramount requirement of technology or product development. Initially, there are technical risks which spell out the difficulties and complexities in developing the technology or realising the product. At production stage, huge investment is needed and scarcity of it causes investment risk. Finally, the economic risk is there when the product comes to the market. Now the investment has to be returned with profit.

Managing Complexities

The challenges of complexities and uncertainties in new technology development are combated by assigning specific roles to specific organisations. Partnerships and consortiums are very useful in this. There should be synergy between development partners, production partners and the user. Sometimes, redundancy for critical technology and products opens up many parallel options such as multiple technology options, multiple sources for the same item, technology collaborations, etc., to realise the goal. The issues related to quality control and its management are very pertinent to combating the complexities in new technology development. Quality assurance and reliability are important factors as the user always expects the best product. Reliability improvement could be done using several techniques like reliability modelling, simulation, predictions and growth. The prudence of visualising the potent problem helps in finding a timely solution. The latest tools available in analysis, modelling and simulations solve many engineering problems pertaining to development. The mechanisms of planning, reviews and monitoring enable us to understand complex situations quite well and also enable us to find remedial solutions.

Strategies for Technology Development

A new technology may be developed based on the correct assessment of its status. If new technology is available off the shelf then outsourcing is the best option. Otherwise, its substitution may be considered. If substitution is not possible, then the agency can be identified for product development. If product development cannot

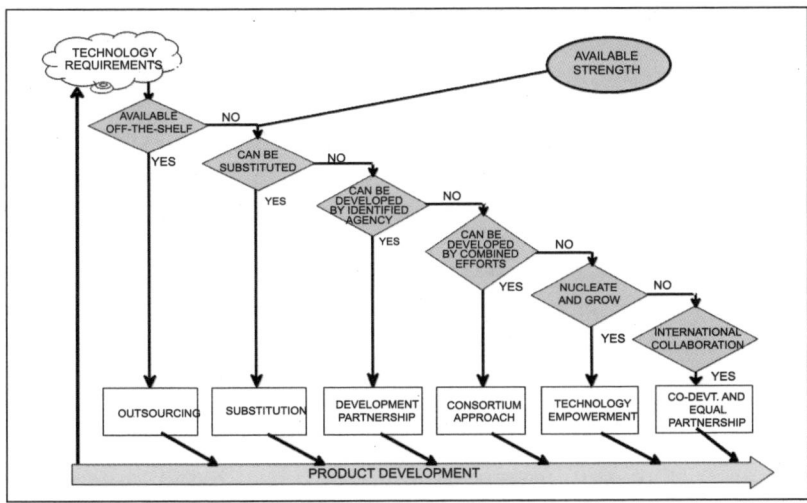

Fig.3.76: Strategy for Technology Development

be affected by any agency, then the next stage is that the development of the product would be done through a consortium of the combined efforts of collaborating partners such as institutions/industries. Possibilities of partnership and consortiums are also methods to develop new technology. Technology empowerment is the final option for indigenous development of new technology. It is a time-dependent process but many new capabilities are developed during the course of the development of new technology. Sometimes, international collaborations and joint ventures help in complementing each other's technologies to produce a world-class product (Figure 3.76).

Case Studies of Technology Development

The guidance and control system of Prithvi was a challenge as it is guided by an on-board computer programme during its terminal phase. This makes it unique and independent of a ground control system and, hence, it cannot be jammed electronically by the enemy. The development of this technology started from scratch from a coarse sensor and it was refined with new guidance algorithm, Inertial Guidance System (IGS) along with On-Board Computer (OBC) package. This could be developed after intensive research by pooling technical resources within the time schedule.

With the coarse sensor, the desired accuracy cannot be achieved and leads to error during the flight of the missile. Error modelling had been done to improve the accuracy of the sensor. Software had been coded for drift compensation by Research Centre Imarat (RCI). A new guidance algorithm was developed by the academia and was verified using simulation. Engineering prototype fabrication was done by the industry and the system was tested using thermal model and mode analysis for high accuracy. Finally, it resulted in the technology capability of highly accurate IGS. R&D activities were carried out at RCI, Hyderabad, and various subsystems were made with the collaboration of the academia and the industry. Later, with the collaboration of HAL, these systems were produced and used in the missile. These systems met the expectations of the mission and the end user. This is an ideal example of technology development using substitution. The same coarse sensor was used by the substitution of software in the sensor to achieve the desired result.

There was a challenging task to develop a nose cone for the re-entry stage of the Agni missile which re-enters at hypersonic speed of ~6 Mach. Re-entry being the most critical aspect of this class of

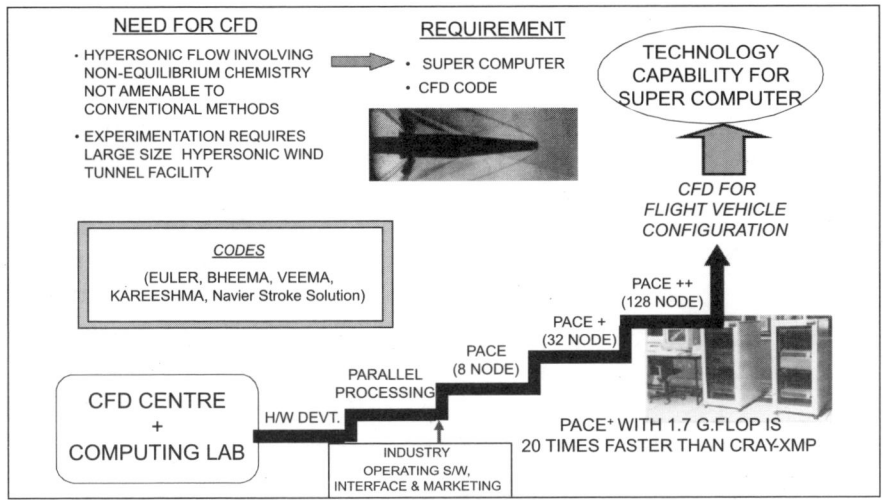

Fig. 3.77: Super Computer for Computational Fluid Dynamics

missiles, we had to configure the optimal design of the re-entry nose cone. This required a hypersonic test facility for evaluating hundreds of configurations which did not exist in India at that time, so there was a requirement of a mathematical solution to the problem. The technique used was Computational Fluid Dynamic and it required supercomputing (Figure 3.77).

India had a CRAY XMP super computer for weather prediction at Indian Meteorological Department, New Delhi. Restrictions were imposed by the US Government not to use this computer for any other purpose. Hence, India could not use it for CFD analysis though it was available. This triggered our scientists to develop within the shortest possible time a parallel processing computing system which can meet the requirements of Agni and similar projects. Initially, the work started using conventional Euler Codes and conventional PCs where each iteration took nearly eight to nine days. Indigenously developed advanced software codes like BHEEMA and KAREESHMA by our young scientists/engineers, mostly women, led to PACE++ 128 node super computer and these iterations now take less than 8 minutes compared to 8-9 days. The CFD centre at the Indian Institute of Science has emerged as a Centre Of Excellence (COE) in India.

Carbon-carbon composite technology came as a solution to the problem of making a nose tip for Agni because of its ability to ablate at high temperatures and velocity. But there were three challenges in the development of this technology and they were 3D weaving and

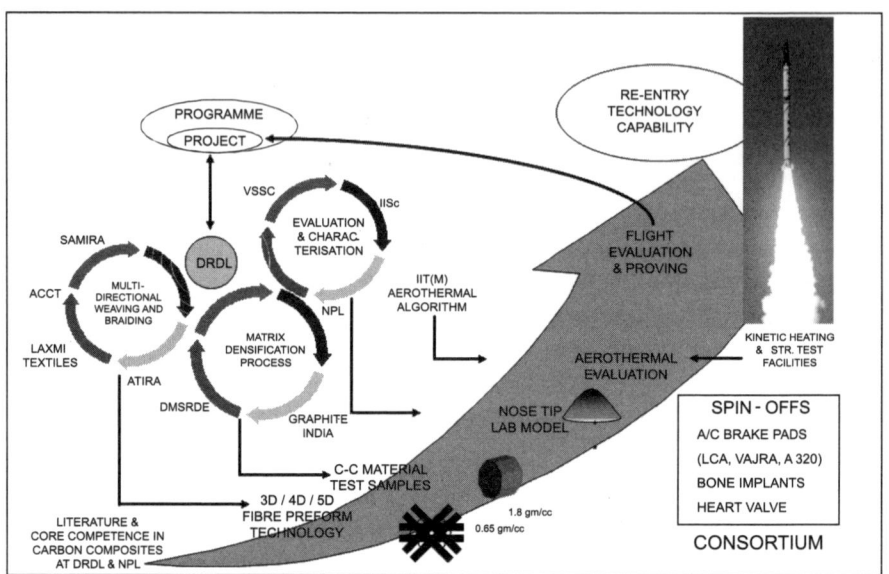

Fig. 3.78: Carbon-carbon Technology

braiding of carbon fibres, densification of matrix and characterisation of carbon-carbon composite. In order to develop these technologies, a consortium was built involving academia (IIT Delhi) and industries, and finally, a nose cone technology was successfully developed based on these technologies (Figure 3.78). It was successfully tried out and tested in the laboratory before it achieved astounding success in the Agni flight trial.

Later, the spin-offs of this technology were used in the brake pad of lightweight aircraft. This technology has tremendous medical potential too; owing to the inertness of carbon it can be used as implants. In the process of making this technology, core competencies of many industries, academia and R&D laboratories were explored and consolidated through consortium approach.

The guidance technology of Nag missiles which is a third-generation antitank top attack missile and works on the fire-and-forget principle threw a new challenge of development of an imaging infrared seeker. This was a new and challenging area of technology where many complexities were involved in its development. When the whole world at that time was thinking of developing the wire guided anti-tank missile, India with her forethought decided to develop an anti tank missile without wire guidance technology and, in its place, the use of imaging infra red seeker for guidance of the missile (Figure 3.79). This, too, was developed by forming a consortium of different

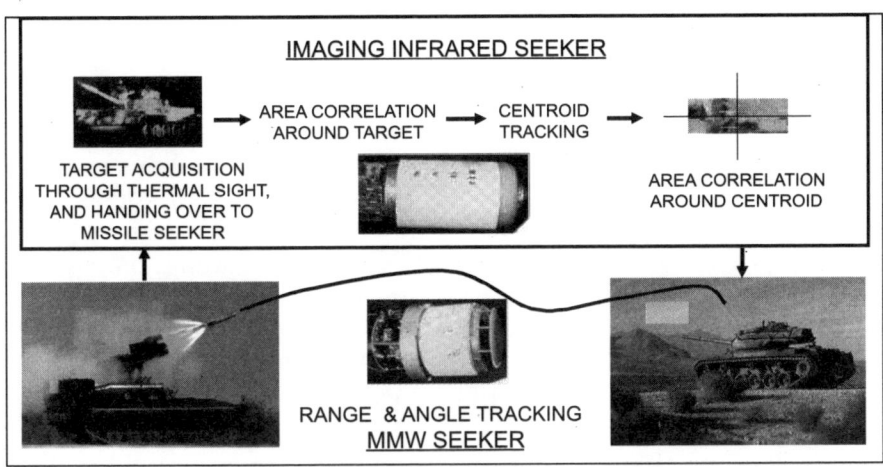

Fig. 3.79: Nag Guidance Technology

groups of DRDO laboratories and academia. Multitasking and apt utilization of resources with constant monitoring resulted in the successful development of an imaging infrared seeker.

Command guidance integrated with phased array radar was required for the multi target surface-to-air Akash missile. Phase shifters, used in the array, were critical and only steered the electromagnetic beam effectively and quickly. Development of phase shifter was a new technology. Phase shifter plays an important role in multi targeting and guidance capability (Figure 3.80). In this process,

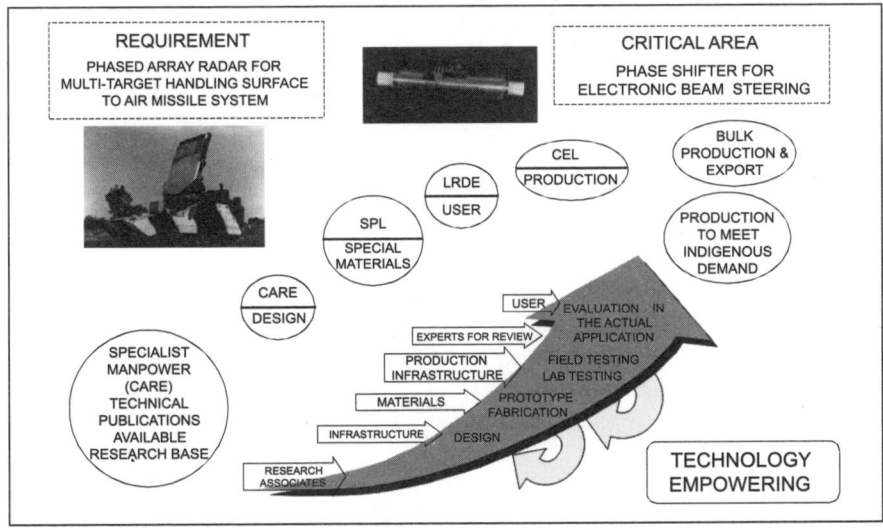

Fig. 3.80: Development of Phase Shifter

ground radar identifies and locks multiple targets. Industry was also involved from its inception for the smooth transfer of technology. Today phase shifter has huge market potential for many applications in radar.

Right assessment of technology strengths and weaknesses is an important factor for the development of high technologies and systems. A systematic approach with fertile brains, assigning the activity, providing the requisite technical and managerial input, formation of consortium, always keeping user in the picture through the development process, quality and reliability inputs, review and monitoring systems in place and finally, a keen involvement of production agency, make the successful path for a globally competitive product. The case studies presented here are the glimpses of mechanisms and methodologies adopted to realise critical technologies. Of course, there are many other areas, where India has attained self-reliance in acquiring competence and proved to have mastered the critical technologies.

Technology Denied is Technology Gained

The Missile Technology Control Regime (MTCR) has denied India major missile technologies and India converted this denial into an opportunity to develop its own missile technologies. During IGMDP, many key technologies were developed indigenously and proven on actual products. Some such technologies are control and guidance, image processing and target acquisition, development of supercomputer, radars for target location, advanced composites, etc. (Figure 3.81). Most importantly, the industries were also developed simultaneously and most of the subsystems are manufactured by industries. The development efforts in propulsion technology and liquid engine technology were remarkable for all missile systems.

The global assessment of IGMDP (Figure 3.82) suggests that performance of all the missiles was well conceived at the time of their design inception stage to perform adequately in contemporary times when they will be ready for induction after a gestation period of development. Today, Prithvi and Agni have become the strategic missiles armed with nuclear warheads giving teeth to the country. Akash, the multi target surface-to-air missile has been inducted in the Air Force and Army. Nag the third generation anti-tank missile will soon get into the Army.

Fig. 3.81: Realisation of Critical Technologies

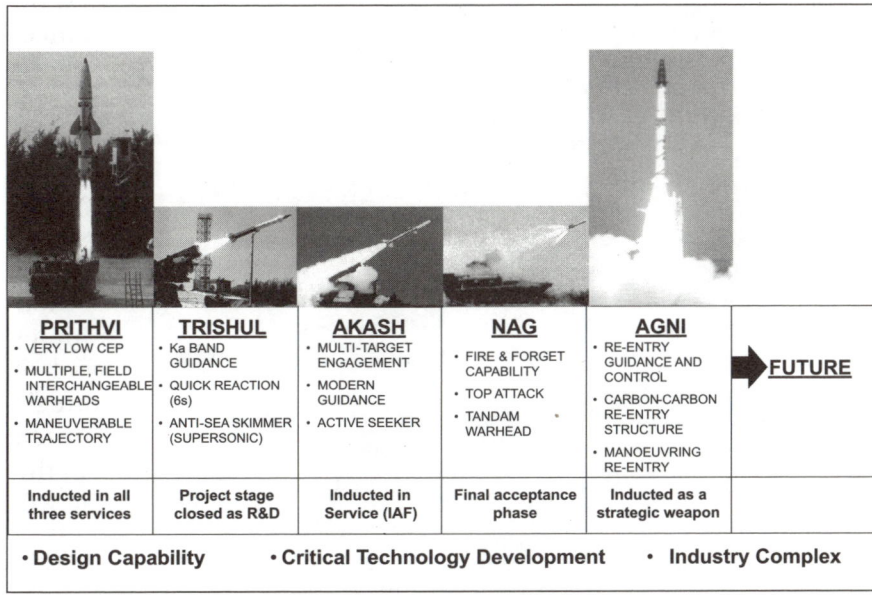

Fig. 3.82: Global Assessment

3.10.3 Ballistic Missile Defence System

The new concept of defending the incoming missile started developing in the minds of the people during the1950s. Scientists were concentrating on developing the technology for detection of launch of ballistic missiles and tracking of the same. Missile defence is divided into categories based on various characteristics: type/range of missile intercepted, the trajectory phase where the intercept occurs, and whether intercepted inside or outside the Earth's atmosphere. Ballistic missiles can be intercepted at any of the trajectory phase - boost, midcourse or terminal. Missile intercept location could be either endo atmospheric or exo atmospheric.

Indian initiative on the Ballistic Missile Defence started in the year 1999. The Ballistic Missile Defence (BMD) system is a prestigious programme of the DRDO and has proven itself with successful flight trials in the recent past. It is a two-tier BMD system, capable of tracking and destroying incoming hostile missiles both inside (endo) and outside (exo) the earth's atmosphere (Figure 3.83). The system consists of two interceptor missiles, namely Prithvi Air Defence (PAD) for high altitude interception and Advanced Air Defence (AAD) for lower altitude interception, radars [Long Range Tracking Radar (LRTR), Multifunction Fire Control Radar (MFCR)], Launch

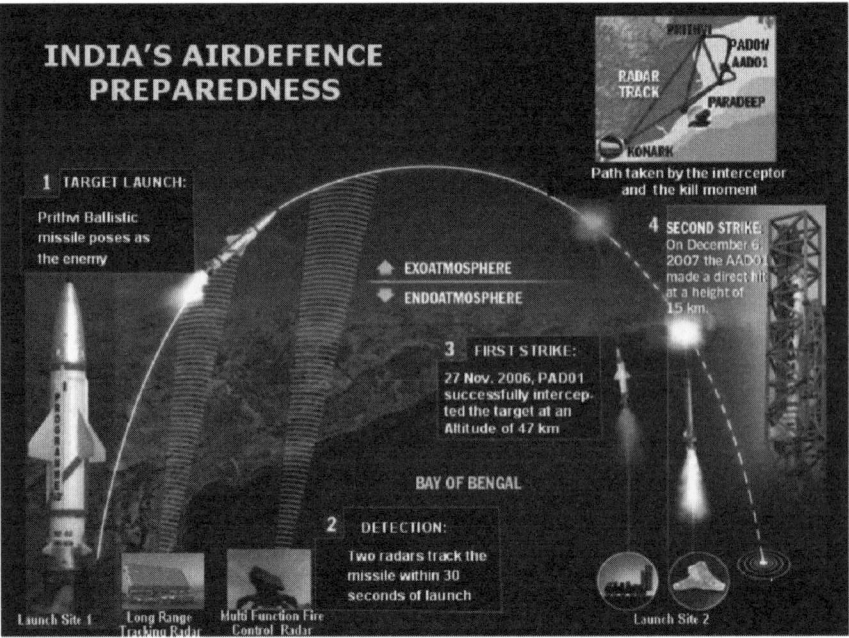

Fig. 3.83: India's Air Defence Preparedness

Control Centres (LCC), Mission Control Centre (MCC), which are geographically distributed and connected through secure communication network. The technologies exclusively required for the ballistic missile defence such as the radars, mission control software, navigation and active radar seeker, etc., have been realised through the various laboratories of DRDO. LRTR, the target acquisition and fire control radar for the BMD system, has the capability to track 200 targets in a range of 600 to 800 km and can spot objects of very small size. MCC–a software intensive system of BMD receives information from radars and satellites that are processed by the computers. In total, MCC carries out all tasks such as target classification, assignment, number of interceptors required to neutralise the target for an assured kill probability and passes on the information to LCC.

Two-stage Interceptor missile PAD with first stage solid fuelled motor and second stage liquid fuelled is capable of intercepting the incoming missile at exo-atmospheric altitudes between 50-80 km. Manoeuvre thrusters generate lateral acceleration of more than 5g at 50 km altitude. Inertial navigation system provides guidance with mid-course updates from the LRTR and active radar homing in the terminal phase. The missile has the capability to engage targets of 300-2000 km class of ballistic missiles. AAD is a single stage solid fuelled missile and capable of destroying the incoming ballistic missile in the endo-atmosphere at altitudes between 15–30 km. Same as PAD, it has an inertial navigation system with mid-course update from ground radar and active radar homing in the terminal phase. With the successful launches of PAD and AAD, India has the capability of full-fledged multi-layer ballistic missile defence.

3.10.4 Cruise Missile - BRAHMOS

The changing theatre of war in the latter part of the twentieth century has led to the evolution of many new weapons. The dominant weapon in the Gulf Wars was the deployment of Tomahawk cruise missiles. More than 1000 Tomahawk missiles were launched at the beginning of the wars precisely to destroy the enemy assets like communication and control centres, power supply, missile and ammunition depots, air fields. The high manoeuvrability, undetectability with stealth features and flying at low altitudes, precision in hitting the targets and launch capability from multiple platforms with stand-off distance made the Tomahawk missiles deadly in the Gulf Wars.

The enormous capability of the cruise missiles seen in the wars,

made India work for the design and development of a cruise missile
initially for application as anti ship missile. The cruise missile is a self-
propelled guided airborne vehicle which has aerodynamic lift due to
its wings and continuous operation propulsion throughout its flight.
Cruise missile technology uses various propulsion systems such as
turbojet (used in aircraft), liquid ramjet and solid ramjet (used in
supersonic missiles) and scramjet (supersonic combustion ramjet
used for hypersonic airborne vehicles). The advantages of cruise
missiles are (a) control of thrust during its flight, (b) operability at a
wide range of altitudes, (c) compactness in size giving minimum radar
cross section signature, (d) excellent fuel economy as the oxidiser is
from atmosphere and (e) high precision. In addition, if it has
supersonic speed, it will give tremendous effect in destruction of the
target, apart from minimum reaction time given to the enemy to
defend himself.

The energy level for each of the propulsion systems is measured in
terms of Specific Impulse (Second). Solid and liquid propellants
provide 300 seconds, Cryogenics give 450 seconds, Solid ramjet gives
600 seconds and the liquid ramjet gives 1200 seconds of specific
impulse along with the possible speed of the system in Mach Number
(Figure 3.84).

Liquid ramjet engine has a definite advantage of four times the
energy level of solid propellants in addition to low weight, low volume,
excellent fuel economy and a wide range of altitude of operation in

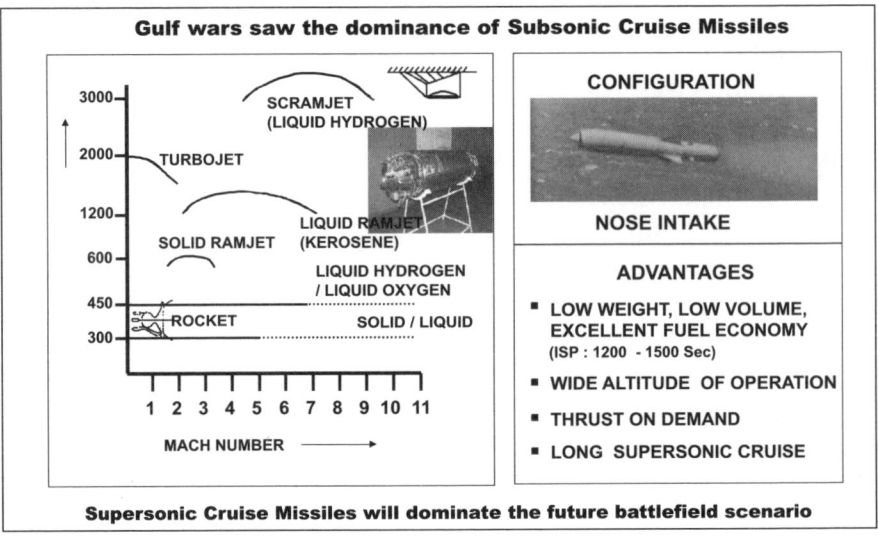

Fig. 3.84: Cruise Missiles – The Ultimate Force Multiplier

supersonic cruise mode and also long time storability. Utilising the availability of the liquid ramjet engine with the NPO Mashinostroyenia, Russia, DRDO decided to work with them to evolve a feasible jointly designed cruise missile. That is BrahMos, named after the two rivers Brahmaputra and Moskva of India and Russia.

BRAHMOS is a canisterised missile and can be launched from ship, land, silo, submarine and aircraft (Figure 3.85). The accuracy and speed of the missile and versatility of launch platforms make it most lethal. It has a first stage of solid propellant and a second stage of liquid ramjet engine. It is guided by inertial navigation during the course of its flight and during the terminal phase it locks on the target with the help of an onboard seeker. It is a tactical missile and has highly manoeuvrable trajectories. It has a range of 290 km and velocity of 2.8 Mach. In sea attack mode it skims over the sea surface and hits the target.

Fig. 3.85: BRAHMOS Supersonic Cruise Missile

The system was then proven from Mobile autonomous launcher suiting the requirement of Indian Army. BRAHMOS Land version has been inducted by Indian Army. Land attack capability of the missile has been proven with the development of BRAHMOS Block II version with target discriminating capability and advanced features. To enhance the capabilities of the missile further, the Block-III version with supersonic manoeuvres and steep dive capability has been successfully carried out. Indian Army is the only land force in the world to have such a formidable supersonic land attack cruise missile for mountain operations. The versions of BRAHMOS are depicted in Figure 3.86. The emergence and induction of BRAHMOS in the Indian Armed Forces speaks that India has an armoury which is second to none with a winning edge as a first strike weapon.

India's decision to go for multi-role supersonic cruise missiles in a JV mode is justified by the fact that India and Russia are the only two countries to have this type of advanced guided missile system in the world. BRAHMOS is the world leader and presently there is no equivalent operational cruise missile in the world in speed, precision and power.

Thus, India has mastered the major critical missile technologies, combating the missile technology control regimes, with the help of multiple academic institutions, R&D labs and industries. India today is self–reliant in design and development of any type of missile system including the Inter Continental Ballistic Missile and submarine launched cruise and ballistic missiles. India is a missile power. The collaborations and joint ventures also provided a new working environment and challenges in overcoming hurdles in the

Fig. 3.86: BRAHMOS Capability Enhancement

development of certain key technologies at par with the developed nations. In fact BRAHMOS has put India ahead of the developed nations, being world number one. The success in these high technology programmes reinstated the faith and confidence among the countrymen that we too can create world-class technologies and systems.

Changing Dimensions of War Theatre

The first 'ballistic weapons' probably were rocks that caveman hurled at each other. War weaponry has gone through various phases of development, particularly during the twentieth century with new weapons and platforms. In the twenty first century, BRAHMOS a supersonic cruise missile, which is more versatile and capable of multi-role, multi-target and multiple missions, has come a long way to intelligent and autonomous systems with faster operations capable of utilising deep sea and space 'networked' sensors and weapons. The pace of future battles will be so swift that there will be no time to revert to rear headquarters for instructions and advice. A typical military hierarchical structure in such an environment will fail. Network centric warfare enables a shift from attrition style warfare to a much

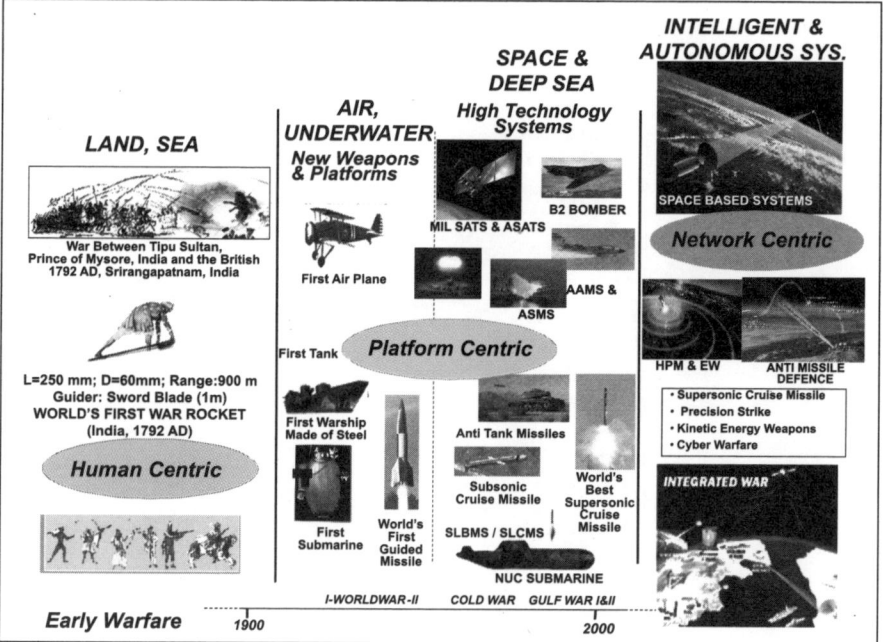

Fig. 3.87: Changing Dimensions of War Theatre

faster and more effective war-fighting style characterised by the new concepts of the speed of command and self synchronisation. Strategically, it allows an understanding of all elements of battle space and battle time; operationally it provides a close linkage between the units and the operating environment, and tactically it provides speed. BrahMos as an independent weapon today fits in as part of the C4I network where the user in land, sea, air and sub-sea would be in a position to use it in strategic and tactical missions. The strategic mission would be for protection of assets from attack while tactical mission would be for the use of operational forces during combat operations.

3.11 HYPERSONICS

3.11.1 Hypersonic Reusable Missile

There are different types of ballistic and cruise missiles with different ranges that have been designed and are available with most of the countries. As it has been described in the earlier part, different propulsion systems are used for different speed regime according to their specific impulse. The different sonic regimes and the platform systems, supersonic biplanes and missiles operated in that regime are described in Figure 3.88. India has already succeeded in developing

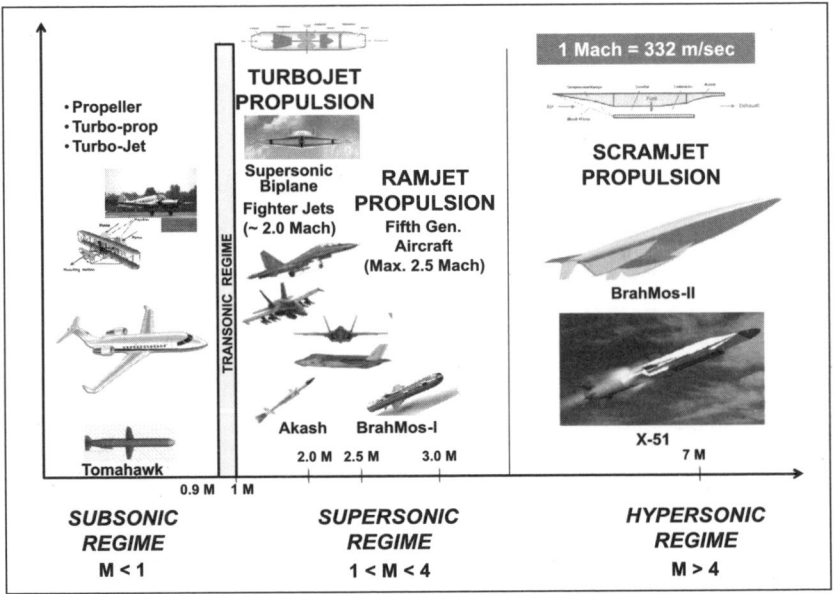

Fig.3.88: Sonic Regimes

and inducting the supersonic cruise missile as a joint venture with Russia. To keep up the lead in the Cruise missile technology, India and Russia together have agreed to progress on the design and development of Hypersonic cruise missile. The future would be the hypersonic regime.

Attempts are being made worldwide in establishing technologies required for realising a missile that can fly in the hypersonic speed of more than 4 M. DRDO is progressing the development of Hypersonic Technology Demonstrator Vehicle (HSTDV) with an aim of demonstrating autonomous flight of a scramjet integrated vehicle using aviation turbine fuel (Figure 3.89). The hypersonic vehicle would fly at 30-35 km altitude in 6.2 to 6.5 M with flight duration of 20 sec initially. To realise the technologies being developed at Defence Research and Development Laboratory (DRDL) are Aero-propulsion integrated configuration, hypersonic air-intake, hot structures, single expansion ramp nozzle, scramjet combustors etc. The technologies to be developed for a long duration flight (~600 sec) are Nose tip, wing and tail leading edges, Ni super alloy / Ti alloy honey comb based Thermal Protection System, protective coating, active cooling of scramjet engine, endothermic fuel, etc.

When we think of Hindu Mythology and Lord Vishnu having a Sudharshan Chakra, its technological marvels come to our mind. The Sudharshan is always in motion at right point finger and is ready to annihilate the enemy and return back with success. Can we think of such a weapon which can fly at a very fast speed, execute the mission and come back and land on our soil? This is called Hypersonic Reusable Missile.

A typical hypersonic air launched missile with reusability flying at an altitude of 30-40 km altitude in cruise mode at Mach 7 is described below (Figure 3.90). The typical mission would be carried out through an air launched missile at an altitude of 10 km and boosted to 3.5 M at 15 km altitude. Kerosene based dual mode ramjet propulsion will take to 5 M at an altitude of 23 km. Scramjet propulsion will take the missile at a cruising speed of 7 M at 35 km altitude. The missile would deliver the payload at the designated target and will fly back to its destination.

Preliminary challenges in the design of Hypersonic Technology have been solved by scientists at DRDL to demonstrate scramjet propulsion technology (Figure 3.91). Scramjet combustor has been designed with Nimonic C263 material with a maximum heat flux of $150\,W/cm^2$ and 1400 K casing temperature. A number of ground tests

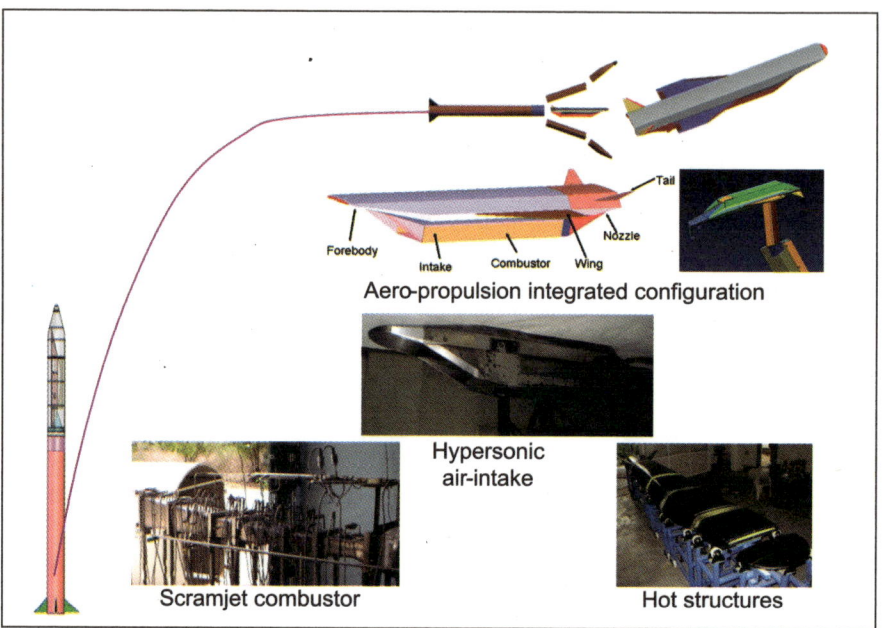

Fig. 3.89: Hypersonic Technology Demonstrator Vehicle

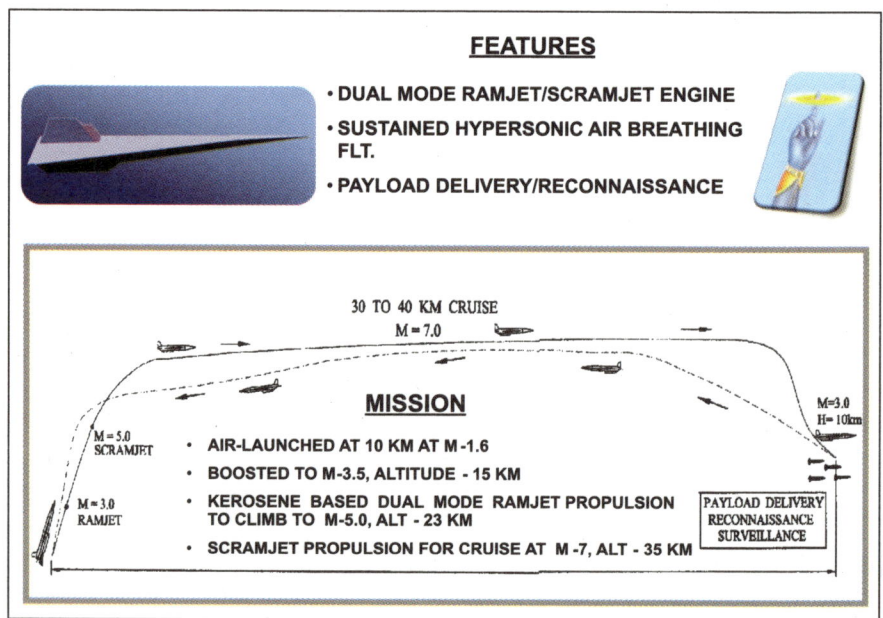

Fig.3.90: Hypersonic Reusable Missile

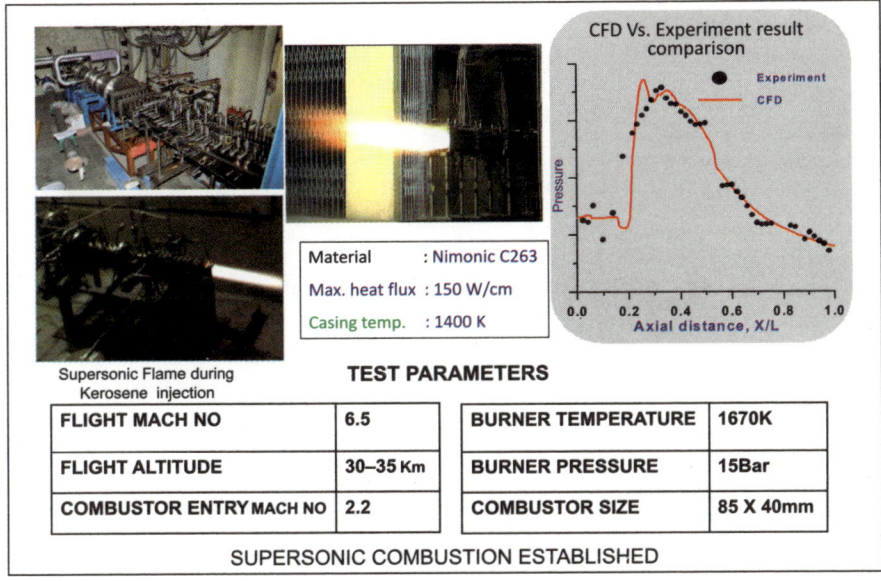

Material	: Nimonic C263
Max. heat flux	: 150 W/cm
Casing temp.	: 1400 K

Supersonic Flame during
Kerosene injection

TEST PARAMETERS

FLIGHT MACH NO	6.5	BURNER TEMPERATURE	1670K
FLIGHT ALTITUDE	30–35 Km	BURNER PRESSURE	15Bar
COMBUSTOR ENTRY MACH NO	2.2	COMBUSTOR SIZE	85 X 40mm

SUPERSONIC COMBUSTION ESTABLISHED

Fig. 3.91: Scramjet Development

has been conducted for 20 seconds duration. Spontaneous ignition and sustained combustion have been achieved under supersonic flow conditions. Performance of the tests have been evaluated using test data and CFD. With this, the supersonic combustion have been established.

Hypersonic Research Centre at IISc

As a consortium of academia, R&D Laboratories, a Memorandum of Understanding for setting up of Centre of Excellence in High Speed Aerodynamics was signed on 12 June 2011 between the BrahMos Joint Venture and the Indian Institute of Science (IISc) for scientific research in advanced technologies including aerospace field. As a first step, Centre of Excellence had been inaugurated on 08 November 2011 in the premises of IISc. The Centre of Excellence established would undertake focused collaborative research in high speed aerodynamics and associated interdisciplinary areas like Hypersonic flow control, New concepts in aerodynamic configuration, Scramjet, Numerical simulation and ground testing, Supersonic gaseous mixing and related complex gas dynamics, Special materials / coating for hypersonic flight, Next generation control and guidance strategies for hypersonic speed   regimes, Special purpose MEMS and

Nano sensor for hypersonic vehicles, Chemical kinetics associated with fuels in Scramjet engines, etc. Likewise, the BrahMos JV has signed a Memorandum of Understanding with the Moscow Aviation Institute (MAI) for the creation of Aerodynamics Research Centre.

The following would be the challenges for the Hypersonic Centre:

* Evolution of CFD Model for hypersonic flight system with the integrated flow studies from subsonic, sonic, supersonic and hypersonic systems
* Hypersonic flow control – Proof of concept experiments in lab to actual applications
* Development of high lift generating surfaces
* Optimum configuration for SCRAMJET powered hypersonic speeds
* Complex gas dynamics at very hypersonic speeds
* Special materials / coatings for hypersonic speed regimes
* Next generation control and guidance for hypersonic vehicles
* Development of special purpose MEMS and Nanosensors
* Understanding of chemical kinetics associated with fuels in SCRAMJET engines

The emerging need is innovation, research and development of technologies for reusable hypersonic flying vehicles. The demand from hypersonic fighter aircraft and hypersonic cruise missiles will be in the areas of hypersonic aerodynamics, propulsions, materials and structures, new CFD codes and so on. This research centre would make India realize many out of box research results of world class status leading to re-usable cruise missile and multiuse single stage to orbit Hyperplane.

3.11.2 Reusable Launch Vehicles

The cost structure of existing launch vehicles is very high because the space vehicles are expendable, i.e., they can be used only once or at best are partially reusable. Therefore, Reusable Launch Vehicles (RLV) can provide a viable alternative for low-cost access to space, applications of space tourism, manufacturing in zero gravity, transportation of cargo for building space colony and new space industrialization (Figure 3.92). In the RLV category, there are two types of vehicles which are being considered, the first being Two Stage to Orbit (TSTO) powered by rocket propulsion and Single Stage to Orbit (SSTO) powered by air breathing propulsion. While both TSTO and SSTO vehicles are fully reusable the same number of times,

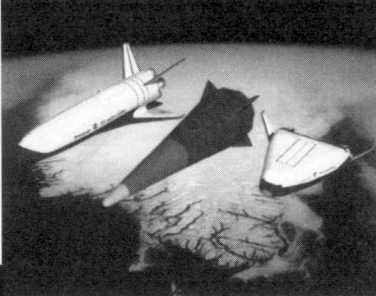

TECHNOLOGIES

- ROCKET PROPULSION (TSTO)
- AIR BREATHING PROPULSION (SSTO)

LAUNCH	COST
TSTO	Rs.1,00,000/kg
SSTO	Rs.10,000/kg

BENEFITS

- SPACE TOURISM
- SPACE MANUFACTURING
- COST EFFECTIVE TRANSPORTATION
- HIGHER PAYLOAD EFFICIENCY

Fig. 3.92: Reusable Launch Vehicle

air breathing heavy lift aerobic SSTO could potentially achieve three to ten times the payload efficiency of rocket propelled TSTOs. This would enable the reduction in cost of access to space if SSTOs were to be used from Rs.100,000 per kg to as low as Rs.10,000 per kg. Whereas advances have been made in rocket propulsion, technology is yet to advance in the area of air breathing propulsion. Needless to say, RLVs would give a boost to space tourism and space manufacturing with cost effective transportation and higher payload efficiency.

3.11.3 Hyperplane

India proposed in the International Astronautical Federation in 1988 at Bangalore, for the first time, a new concept of Hyperplane. Lead taken by Air Cmde Gopalaswamy, the team brought out that the payload efficiency can be improved to 15 per cent compared to the 7 to 8 per cent proposed by USA, Europe, Japan. The idea is to develop a hyperplane vehicle that can take off from conventional airfields, collect air in the atmosphere on the way up, liquefy it, separate oxygen and store it on board for subsequent flight beyond the atmosphere. It would take off horizontally like a conventional airplane from a conventional airstrip using turbo-ramjet engines that burn air and hydrogen. Once at a cruising altitude, the vehicle would use scramjet air breathing propulsion to accelerate from Mach 4 to Mach 8. During

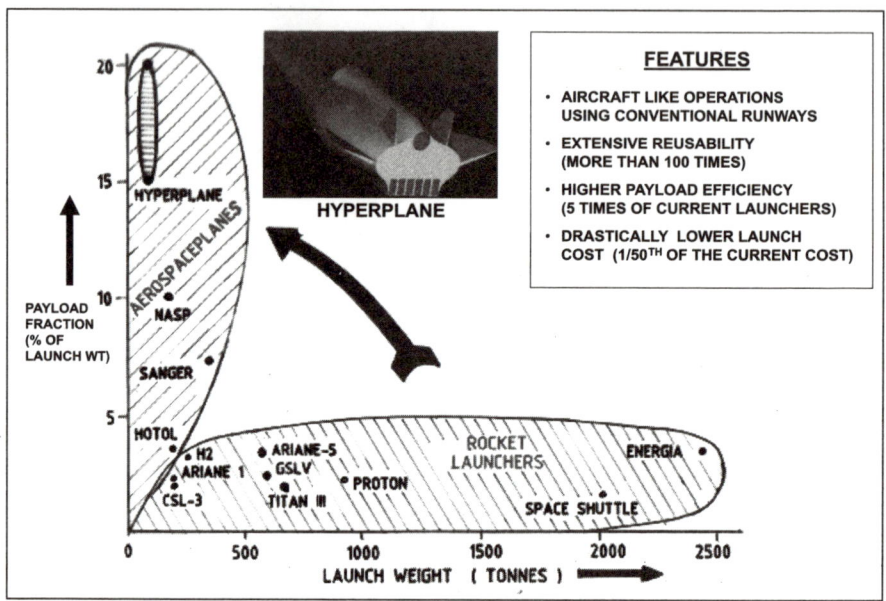

Fig. 3.93: Multi-Purpose Aerospace Vehicle

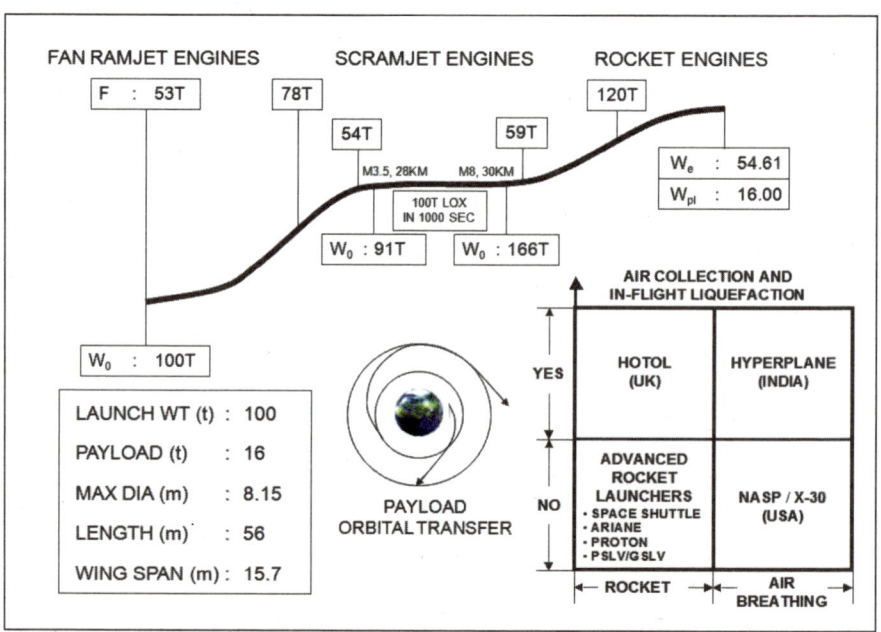

Fig. 3.94: Hyperplane Mission Profile

this cruising phase, an on-board heat exchange would collect the hot air from the engine and convert it into liquid oxygen. The liquid oxygen collected would then be used in the final flight phase when the rocket engine burns the collected liquid oxygen and the carried hydrogen to attain orbit. The vehicle would be designed to permit at least a hundred re-entries into the atmosphere. When operational, it is planned to be capable of delivering a payload weighing up to 1,000 kg to low earth orbit. It would be the cheapest way to deliver material to space.

This type of mission will be highly useful for multiple applications. In the case of Hyperplane, the aim was to achieve larger payload fraction. The space shuttle of USA with 2,000 tonnes takeoff weight could launch only 30 tonnes in low earth orbit, giving a payload fraction of 1.5 per cent. India's concept of Hyperplane aims to realise 15 per cent of payload fraction (Figure 3.93). This will considerably reduce the launch cost per mission and will enable multiple missions such as transport, reconnaissance, payload delivery, satellite injection, etc.

On a typical mission, Hyperplane would take off with 100 tonnes weight using fan ramjet engine and then on scramjet mode for nearly 1000 sec. during which time it collects the left over air, cools it and separates as liquid oxygen. This increases its weight to 166 tonnes, thereafter it flies in rocket engine mode using the liquid oxygen and stored liquid hydrogen to deliver a payload of 16 tonnes (Figure 3.94). This concept of mass addition in flight is unique and has been conceived by Indian scientists.

3.11.4 Low Cost Access to Space

A cost analysis for access to space for different space missions of launching satellites using conventional expendable launch vehicles, orbital transportation by partial reusable vehicles like space shuttle and large missions such as space colonisation using fully reusable Hyperplane are brought out in Figure 3.95. The price per tonne for the low earth orbit can be drastically reduced from $10 million to $1000 by going for the Hyperplane approach.

Critical Technologies for Space Transportation Systems

The real value of future societal space missions, like energy from space and seawater desalination using space solar power can take place only when mankind builds fully reusable space transportation

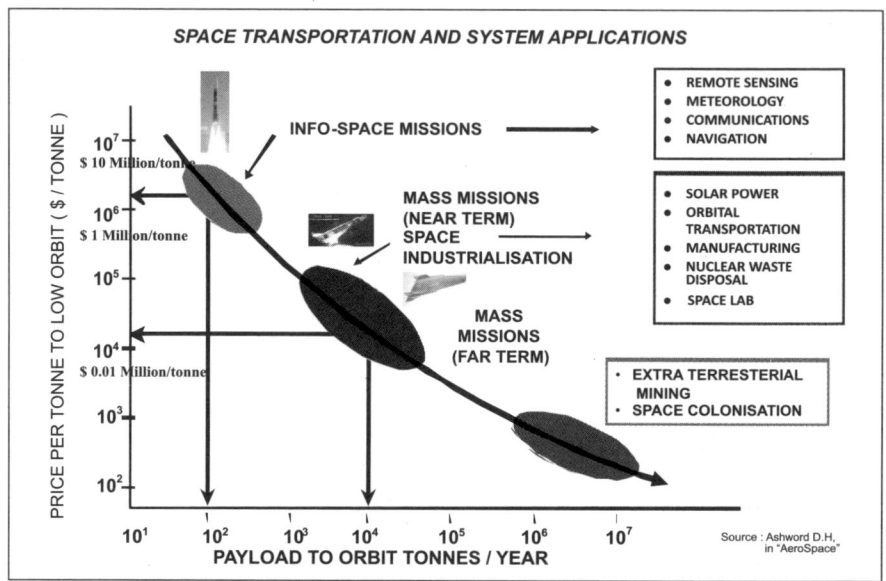

Fig. 3.95: Cost of Access to Space

systems with very high payload efficiencies. Several technologies are critical and need to be developed such as:

- In-Flight Air Collection and Oxygen Liquefaction Technology
- Ram/Scramjet Engines
- Ascent Turbojet /Turbofan Ramjet Engines
- Advanced light weight High Temperature Materials

3.12 EMERGING BATTLEFIELD TECHNOLOGIES-C4I2SR

The developments in information technologies together with satellite technology and other technologies like smart materials, propulsion systems, etc. will have an effect on warfare. The battlefield scenario of the twenty first century is characterized by high technology weapons and systems on one side and a defused battlefield on the other. The technologies of digital battlefield, C4I2SR networks, electronic warfare and information warfare will transform the nature of war. At the same time, the world is faced with new regional capabilities, low intensity conflicts, proxy wars and sponsored terrorism. As a result, the old paradigm of military strength being determined by numerical superiority, sustained capability to fight and capability to inflict maximum damage is giving way to a new paradigm of weapon effectiveness and force mix rapid response to threat and precision strike capability. New capabilities of anti-missile defences, beamed energy weapons and non-lethal weapons are emerging and the battlefield is extending to the geo orbit, deep oceans and to cyberspace. Future missile systems will have to operate in this environment in an integrated manner. It is the combination of platforms, weapons, sensors, command and control, and the information technology systems that will give the required force multiplication for future weapon systems.

Technology will dominate physical warfare with unmanned warfare taking centre stage. Miniaturisation with the development of MEMS and Nanotechnology will be widely applied. Future warfare will see the development of tactical Unmanned Aerial Vehicles (UAV), Micro unmanned aerial vehicles, unmanned underwater vehicles, unmanned ground vehicles, unmanned combat air vehicles, stealth ships to name only a few. Unmanned air vehicles are capable of flying at high altitude and firing missiles. Micro UAVs can be used by a soldier to give him a series of information. Unmanned underwater vehicles are capable of going deep into the water to study the enemy position and fire at targets. Unmanned ground vehicles can detect and cross mines and fire by remote control, thus, protecting soldiers. Similarly, for unmanned combat vehicles, men will not be directly involved.

Therefore, on account of these developments in technology, men will be operating from behind the scene of action. In the area of missile systems, the three most essential characteristics of the twenty first century could be summarized as zero CEP precision, hypersonic speed and reusability. Hypersonic speed will result in maximum flight time, low detectability, and low interceptability. Because of the minimum reaction time available to the target defences, the hypersonic missiles can penetrate even heavily defended targets. Reusability will drastically reduce the operational cost of the missile systems and also open up multiple applications for missiles including reconnaissance, payload delivery and damage assessment.

Another promising futuristic development will be that of the Smart Soldier. The smart soldier will have an integrated helmet having computer and sensor display, night vision instruments, communication systems, etc. He will have adequate body protection equipped with smart clothing including smart shoes, which will have ballistic protection at a reduced weight, mine sensors, automatic racing technology to enable the soldier to walk fast, etc. He will have weapon instrumentation comprising thermal weapon sight, laser range finder, digital compass, GPS receiver, etc. Thus, it is technology more than anything else which will have a profound effect in the battlefield scenario of the future.

3.12.1 C4I2SR

Command, Control, Communication, Computer, Information, Intelligence, Surveillance and Reconnaissance—C4I2SR or C4I2SR—is becoming a vital component in battle preparedness for tomorrow. In the new paradigm of a war theatre, integrated and networked decision support system with space based, land based, sea based and undersea sensors, weapons and equipment will be dominating for effective and optimum utilization of resources to inflict maximum damage to the enemy. The Changing Dimensions of War Theatre, C4I2SR in Ballistic Missile Defence, Tactical C4I System, Universal BrahMos in Network Centric Warfare, Global Net Centric Surveillance and Targeting, Future Opportunities for Enhancing ISR and Integrated Ocean, etc., C4I2SR have been detailed in the following paragraphs.

3.12.2 C4I2SR in Ballistic Missile Defence

With the proliferation of missiles and nuclear weapons, today's world is threatened. The Ballistic Missile Defence System (BMD) serves as a forward-deployed sensor by extending the battle space and providing early warning of an intercontinental ballistic missile launch. Sensors transmit track data to the Ground-Based Midcourse Defence command centre via the BMDS. The Long-Range Surveillance and Track capability assists in the defence by providing tracking data to cue other system sensors and initiate a Ground-Based Midcourse Defence engagement. The Command and Control, Battle Management, and Communications element is the backbone of integrated, layered Ballistic Missile Defence and is the nervous system of the BMDS. It ensures the critical flow of information for the survival of the nation and its allies. It provides war-fighters at both the strategic and tactical levels of command with the capability to plan and fight with ballistic missile defence, while concurrently tracking all potential ballistic missile threats; directing weapons to engage on a distributed network; and pairing any sensor with the best available weapon system to defeat ballistic missile threats at any range, in any phase of fight in all theatres.

3.12.3 Tactical C4I System

Tactical C4I systems provide battlefield information to the commander to make decisions and control military forces. This will provide comprehensive information expeditiously to troops on the ground. In order to meet the complex operational requirements of today, C4I system should provide interoperability with other systems so that enhanced operation capabilities are achieved with synergy. The Indian Artillery's tactical C4I systems encompass wide area network covering command information decision support; EW and Battle Surveillance; Air Defence; and Air Space Control Systems. The entire network web runs from corps to division to brigade and finally to artillery field deployment units. This extension of tactical C4I system from battalion level down to individual soldier level is the key to the entire communication and data network system. Network centric warfare is a marked departure from conventional warfare and greater amount of transformation would be required in the armed forces to support the new concept of operation.

3.12.4 Global Net Centric Surveillance and Targeting (GNCST)

GNCST is a prototyping effort with the effective use of space, air, land, sea and sub-sea domains in future war fighting capability. Joint Targeting and Attack Assessment by Army, Navy and Air Force as a coherent joint effort would play a major role in defining this. It may also feature automated image processing at optical, infrared, and SAR wavelengths to allow the cueing of image analysts to make a final decision. Finally, it features an automatic target and strike-asset pairing decision aid for timely assignment of air strike and de-confliction of the surrounding battle space. Here, the network of distributed autonomous underwater sensors has the advantages of large area coverage, covert operation, and tolerance of individual node failures. Such a sensor network allows passive acoustic surveillance, distributed active surveillance, and multi-static operation with other collection assets to counter such threats as air-independent diesel submarines. These critical surveillance applications can provide high-leverage knowledge that acts as a force multiplier for both defensive and offensive missions. Today, there are three competing approaches to achieving ultra long-endurance persistent surveillance: satellites; high-altitude and low-endurance unmanned aerial vehicles and high-altitude airships. The measure of any wide-area surveillance system is its ability to survey a large region quickly and to uncover clues to an adversary's forces and intent.

3.12.5 Future Opportunities for Enhancing Intelligence, Surveillance and Reconnaissance (ISR)

Closed-Loop Tasking, Collection, and Exploitation present a vision for future ISR tasking and exploitation based on the view that a system that carries out tasking, collection, integration, interpretation, and exploitation should function as a closed-loop process to provide ISR information needed to support a commander's intent. Central to the vision is providing commanders with a stronger ability to control ISR sensors, tools to assess the adequacy of that commander's ISR picture, and fused/multi-source data. The key is the use of space technology to convert large volumes of space-based and airborne multi-sensor data efficiently into actionable information for tactical commanders. As envisioned, the data would provide the automation required to translate the relevant data from a sensor perspective to a tactical perspective—that is, to a map-based view of all objects in the battle space.

3.12.6 Integrated Ocean C4I2SR

The Indian peninsula covers nearly 70 per cent of ocean frontiers towards the east, west and south. The current security strategies highlight the importance of littoral warfare more than bluewater operations. Hence, it can be emphasized that there is a pressing need for a Technologically Advanced Integrated Surveillance capability over the oceans around us to take effective countermeasures against maritime threats. A well-designed surveillance system must have Multi-sensor Intelligent Architecture functionality and may consist of sensor systems such as sonobuoys and moored-buoys networked as seabed arrays. The important factors in the development of Undersea Sensors Network Technologies are the fusion of various sensing devices with Advanced Digital Processing and networking them with Command and Control. As part of Integrated Surveillance, data would be collated and analyzed from the entire sensor network in order to locate targets over a wide area of coverage. Linking of these kinds of similar networks deployed along the coast of the country will essentially result in an integrated coastal surveillance system.

3.12.7 Next Generation C4I2SR

Future war theatre will be C4I2SR network with interoperability, heterogeneous information content, newer technologies such as Photonics, Stealth, Lasers, Robotics and Artificial Intelligence, NEMS based sensors and highly agile and secure communication in all medium. All these new generation technologies and their integration will pave the way for a new paradigm shift in C4I2SR networked operations (Figure 3.96).

REQUIREMENTS	TECHNOLOGIES
• Network Distributed in "space" and Multiple Domains	• Smart material based sensors • HF Agile Commn. Equipment including U/W Commn.
	• Secure Cloud computing
• Interoperability with Heterogeneous Information Content	• Robotics & AI for surveillance & reconnaissance • Use of Cognitive Sciences • Invisible platforms & weapon sites
• Multi sensor surveillance and multi-source real time information system	• Hypersonic stand off precision missiles

Fig. 3.96: Next Generation C4I2SR

The sea may bring together like-minded countries to form a common pool of security network to help and influence the region's strategic perceptions. Indeed, an old saying is: "The sea unites while the land divides". The economic importance of Malacca and the increasing security needs of this strait are of paramount importance. Thus advanced initiatives in C4I2SR can play a most crucial role in providing the necessary blanket to counter many such threats.

3.12.8 Universal BRAHMOS in Network Centric Warfare

Armed forces the world over are facing a paradoxical situation where they need to fulfil their tasks with decreased resources and decreased manpower. This necessitates working smarter and looking for force multipliers. Network-centric warfare enables us to manage this paradox. Network centric computing is governed by Metcalfe's Law which asserts that the 'Power' of a network is proportional to the square of the number of nodes in the network. It was as early as February 1998, DRDO of India and NPO Mashinostroyenia of Russia decided to team up for joint design and development of a supersonic cruise missile which would be a world class product not available

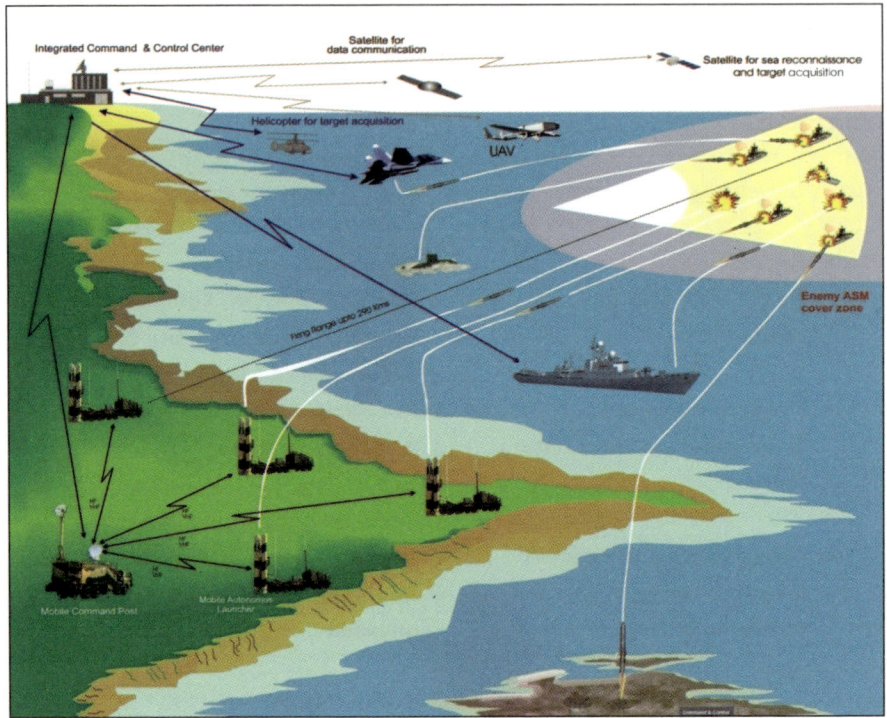

Fig. 3.97: BRAHMOS in Network Centric Warfare

anywhere in the world. Today, this weapon system uses all the cutting edge technology concepts which encompass all the key elements that make this system most compatible in its role of network centric warfare.

Important key technology elements are as follows: Long flight range with supersonic speed all through the flight; shorter flight time leading to lower target dispersion and quicker engagement; wide range of flight trajectories; fire-and-forget principle of operation; higher destructive capability reinforced by the large kinetic energy of impact. In the modern theatre of operations, BRAHMOS has the capability of being launched from land, sea, air and subsea platforms with the highest degree of precision. Common command and control would be able to effectively use this missile system using the latest satellite based communication technologies (Figure 3.97).

Today, there is a paradigm shift from platform centric to network centric warfare which is leading to joint operations and involves a very high level of communication and coordination. BRAHMOS, as a sophisticated weapon system, becomes an integral part of the modern battlefield of tomorrow.

From the above defence technology features, we can realize the future battlefield will be of different class where speed, precision of delivery, destruction of the specific targets and their real time damage assessment with full battlefield knowledge involving space based C4I and unmanned systems and non-lethal weapons. The trend shows the dominance of space based surveillance, use of satellites and Directed Energy Weapons in space, unmanned aerial vehicles using high power microwave and control of enemy's operations through cyber attack. Long range hypersonic precision delivery systems and stealth fighters and submarines will emerge as very important systems for the future war.

3.13 GREEN TECHNOLOGIES

Technological revolution through rapid progress of industrialization left a mark of ecological imbalances. The environment has got its own balance with trees and forestation balancing the CO_2 and oxygen levels. The biggest threat of today is Global Warming. Global Warming is the continuous rise in the average temperature of Earth's atmosphere and oceans. Global warming is caused by increased concentrations of greenhouse gases in the atmosphere resulting from human activities such as deforestation and burning of fossil fuels etc. Awareness has come to protect Mother Earth through

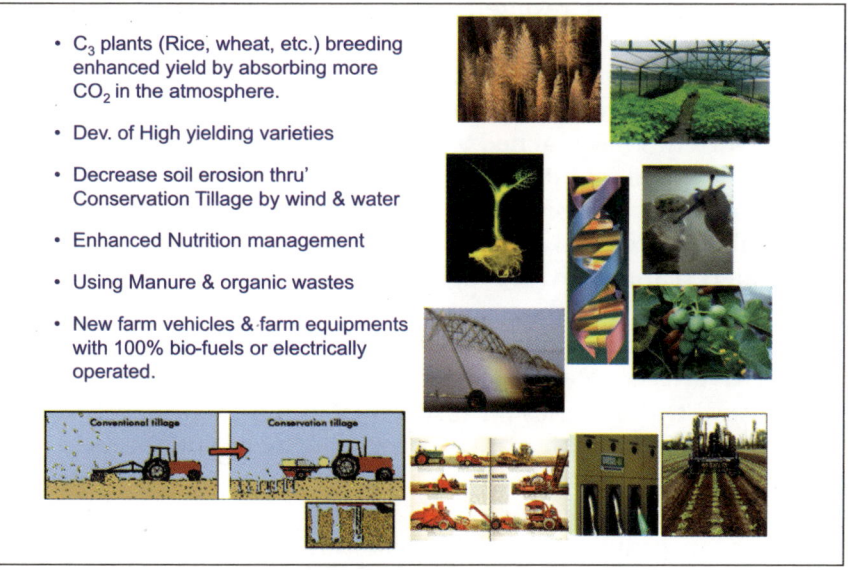

- C_3 plants (Rice, wheat, etc.) breeding enhanced yield by absorbing more CO_2 in the atmosphere.

- Dev. of High yielding varieties

- Decrease soil erosion thru' Conservation Tillage by wind & water

- Enhanced Nutrition management

- Using Manure & organic wastes

- New farm vehicles & farm equipments with 100% bio-fuels or electrically operated.

Fig. 3.98: Green Technologies

sustainable development using green technology.

Green technology or Clean technology is an environmental friendly process used in a way that conserves natural resources and the environment. While Nature's resources and the ecosystem services are declining, the human demand for such resources is increasing. Green technology is to find ways to produce in such a manner that the Earth's natural resources do not get damaged or depleted. Reusability and recycling of the natural resources are given more importance in maintaining the ecosystem. The main objectives of the Green technology are sustainability by meeting the needs of the society without damaging the natural resources, recycle and reuse, reducing waste and pollution by changing patterns of production and consumption. Society, environment and economy are the three pillars of sustainable development. Sustainable development includes a variety of development schemes in social, cleantech (clean energy, clean water and sustainable agriculture) and human resources segments. The world is in transition from the concept of fossil powered energy generation to clean energy such as solar power. The use of nanotechnology based photovoltaic cells for storing of energy, cyclical production through natural resources instead of making and wasting, increased biological diversity etc.

3.13.1 Green Technologies for Agriculture

Global warming has resulted in increased CO_2 percentage in the atmosphere. It favours C_3 plants (Rice, wheat, etc.) breeding which has to be oriented towards enhanced yield by absorbing more CO_2 in the atmosphere and increasing the yield. The productivity of the current varieties has reached a plateau. High yielding varieties need to be developed in addition to higher yield with more quality characteristics viz., more Vitamins and Minerals which will help to reduce the nutritional dependent health problems. Use of micro propagated plants provides high yield capability and is disease free. Radio isotopes help in diagnosing nutritional deficiency in plants and soils for precise application of fertilizers. Irradiation of seeds results in high yield variety cereals.

Assistive technologies for farms such as design farm vehicles and farm equipments which can be run on 100 per cent bio-fuels or are electrically operated help in taking care of environmental consideration. Development of intelligent farm machines like harvester recording the yield per unit area while harvesting, intelligent mechanization like weeding only weeds and not the plants will definitely help the farmers with reduced dependence on labourers and increased management of the yield.

3.13.2 Climate Engineering

Reduction of CO_2 in Air - Carbon Sequestration

Heat from Earth is trapped in the atmosphere due to high levels of Co_2 prohibiting it from releasing into space ("Greenhouse effect"). About half of the greenhouse effect is caused by CO_2. Trees are the best means of reducing the impact of carbon dioxide. Trees remove (sequester) CO_2 from the atmosphere during photosynthesis & return oxygen back to the atmosphere. A single mature tree absorbs CO_2 at the rate of 22 kgs / day and releases enough oxygen back into the atmosphere to support four human beings. Hence, one tree in the city will also save fossil fuel, cutting CO_2 build-up as much as that of fifteen forest trees. Therefore, planting trees is the most effective means of drawing excess CO_2 from the atmosphere. Moreover, planting of trees in the fields protects the crops against wind and sandstorms and leaves shed by the trees to fertilize the soil and improve the microclimate that becomes the most important factor for better harvest.

All the geo-engineering projects have been undertaken worldwide and almost all research has consisted of computer modelling or laboratory tests. Tree planting and cool roof projects are already underway, and ocean iron fertilization is at the initial stage of research with small-scale research trials and global modelling having been completed. The other form is the capture of carbon dioxide and its storage in geological formations, or in marine waters. Solar radiation management reduces the net incoming short-wave (ultra-violet and visible) solar radiation received, by deflecting sunlight, or by increasing the reflectivity of the atmosphere. Adding more fine particles to the stratosphere can increase the sunlight reflecting back into space. Large past volcanic eruptions of Mount Pinatubo produced global scale cooling of about 0.5°C. Obstructing solar radiation with space-based mirrors or other structures like giant reflectors to reflect back the radiation to the space is another form of solar radiation management.

3.13.3 Power and Fuel from Plastic Wastes

Any Material which is not needed by the owner, producer or processor is waste. Waste could be solid waste, liquid waste, plastic waste or metal

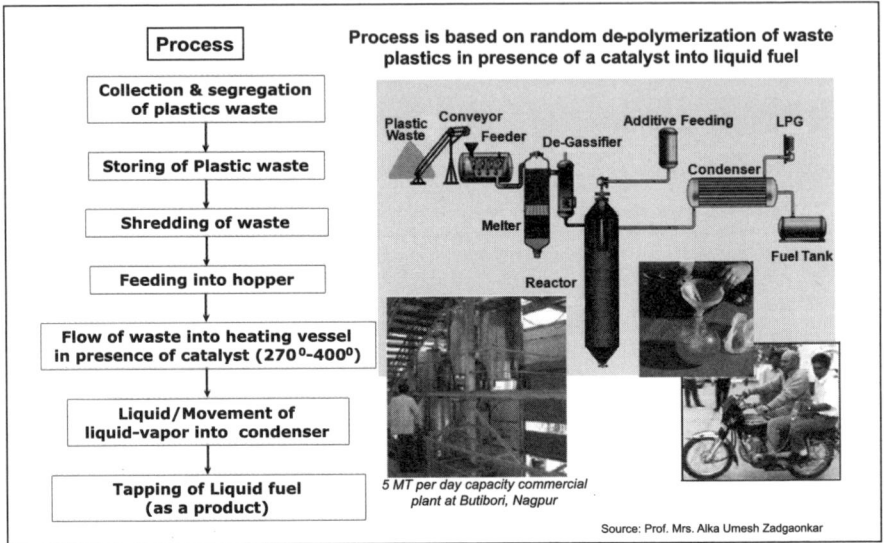

Fig. 3.99: Fuel From Plastic Wastes

waste. The idea is to convert the waste into wealth through technology. Attempts are being made to generate fuel from plastic wastes. Plastic wastes are being converted into fuel through Plasma Pyrolysis Technology (PPT). The Process is based on random de-polymerization of waste plastics in the presence of a catalyst into liquid fuel. Commercial plant is established at Butibori, Nagpur with the capacity of producing 5 MT per day.

Plastic Tar Road

India has 33.4 lakh km of roads (NH & State Highways & Others). Plastic wastes could also be used in making Plastic Tar Road. The bitumen & gravel mix used for laying roads could be combined with flakes or granules made from domestic plastic wastes like carry bags, teacups and a variety of domestic plastics. The advantages are Good skid resistance & texture values, reasonably strong with good even surface, no potholes, rutting, ravelling or edge flaw, for a longer period, and higher binding strength of mix. Salem, an industrial town, is the first city to lay a plastic-tar road in the country.

3.13.4 Power through Solid Waste

The era of wood and bio-mass is almost nearing its end. The age of oil and natural gas would soon be over within the next few decades. The world energy forum has predicted that fossil based oil, coal and gas

Bitumen & gravel mix used for laying roads is combined with flakes or granules made from domestic plastic wastes like carry bags, teacups and variety of domestic plastics.

ADVANTAGES

* **Good skid resistance & texture values.**
* **Entire stretches are reasonably strong.**
* **Good surface evenness.**
* **No potholes, rutting, raveling or edge flaw, for a longer period.**
* **Higher binding strength of mix.**

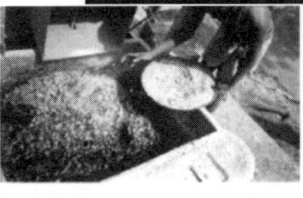

India has 33.4 lakh km of roads (NH & State Highways & Others)

Salem, an industrial town is the first to lay a plastic tar road in the country

Fig. 3.100: Plastic Tar Road

Two plants are operational in India

MSW PROCESSING PLANT

REFUSE DERIVED FUEL (RDF) - PELLETS

PELLETISATION - PROCESS FLOW CHART

GARBAGE

DRYING IN GREEN HOUSE

SEPARATION

SIZE REDUCTION

DENSIFICATION/PELLETISATION

POWER PLANT

* **MSW converted to RDF Coke**
* **Heterogeneous garbage converted to Electricity.**
* **More than 5 Lakh tonnes of MSW being processed**
* **Capacity : 6.6 MW**
* **Exportable : 5.9 MW**

Possible to generate ~5800 MW of Power thru' 900 Plants throughout India

Fig. 3.101: Power through Municipal Solid Waste

reserves will last for another five to ten decades only. The challenge of reducing our dependence on fossil fuels and conserving the environment is a global challenge which would necessarily require evolving and implementing solutions which are efficient and economical. Energy is being generated from the Municipal solid waste (MSW). M/s. Selco International Limited, Hyderabad, the first Project of its kind in India has established a pilot plant with the capacity of 6.6 MW. More than 5 lakh tonnes of MSW is being processed to convert into Refuse Derived Fuel (RDF) Pellets and the heterogeneous garbage converted to Electricity. The alternate sources of clean energy are solar, ocean (Tidal, wave, current and thermal), Wind etc.

3.13.5 Liquid Waste Management

Benefits of liquid waste management through waste water treatment include Lakes & receiving water bodies that would be clean and pristine, treated sewage water can be reused for Irrigation, recycled waste water reduces fresh water demand, environment and sanitation of surroundings improves greatly, biological sludge could be used as manure.

The large presence of dyes (AZO) in effluents from the dyeing and bleaching units causes water contamination. This is a major environmental concern. Possible solution for treatment of water

Benefits of Waste water treatment

- Lakes & receiving water bodies would be clean and pristine
- Treated sewage water reused for Irrigation
- Recycled waste water reduces fresh water demand.
- Environmental and Sanitation of surroundings improved greatly.
- Biological sludge used as manure

Fig. 3.102: Liquid Waste Management

contamination is through enhancing the existing effluent plant through reverse osmosis or switching over to cleaner production technologies such as Advanced Oxidation Process. Centre for Fire, Explosive and Environment Safety (CFEES) of DRDO has worked out on an Advanced oxidation process (AOP) technology. In this process, a combination of UV light, hydrogen peroxide and ozone gas are used to facilitate the oxidation process. By AOP process, a very reactive free radical (hydroxyl radical) is produced in situ the reactor system to destroy the organic contaminants. The hydroxyl radical formed by AOP process increases the rate of reaction to over 100 to 1000 times higher than that observed with either oxidants or UV applied separately. As a result many organic compounds which are refractory and normally resistant to powerful oxidants are destroyed by the AOP in a short time overcoming the inherent shortcomings of the conventional treatment processes. The main advantage of AOP is that it does not produce any secondary toxic byproduct and since the end products are CO_2, water and other non toxic compounds, the treated effluents can be recycled. A pilot plant of 200 LPH based on lab scale data was designed, fabricated and set up at Chennai with the collaboration of industry partner. Four different types of industrial wastes were treated in AOP pilot plant of 200 LPH. From the results it

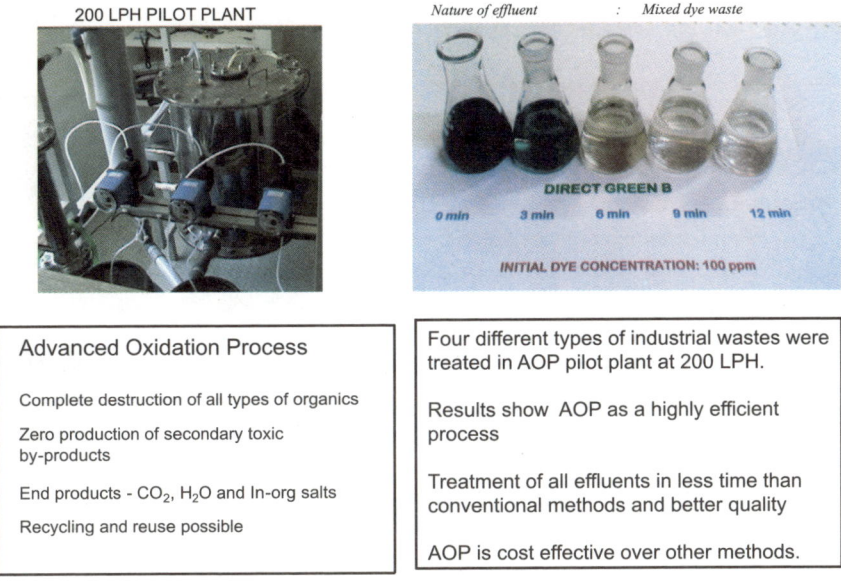

Fig. 3.103: Effluent Treatment Trial–AOP

is clear that AOP is a highly efficient process and can treat all types of effluents in less time than conventional methods with better quality of discharge effluent meeting the CPCB effluent discharge limits. AOP is also cost effective over other methods.

3.13.6 Power through Methane Gas

Fossil fuel emits gas that causes global warming. There is an urgent requirement for the development of alternative energy technologies. Production of Synthetic Natural Gas (SNG) is one of the promising alternatives to petroleum due to its potential as a transportation fuel. Efforts are on to produce SNG i.e. Methane. Methane is produced in a specially built reactor by fusing carbon dioxide and hydrogen. Carbon dioxide is obtained from the exhaust, from fossil fuel, or from chemical fuel like ethanol. Water is dissociated into its elemental components, hydrogen and oxygen, through electrolysis method. Hydrogen is blended with carbon dioxide and the mixture is passed through a special methane reactor. The catalyst-driven reaction produces methane that is chemically equivalent to natural gas. Methane gas is an ideal hydrocarbon fuel with minimal environmental impact as it consumes nearly fifty-seven tonnes of CO_2 to produce one million cubic feet of methane. It returns to CO_2 and

AN IDEAL HYDROCARBON FUEL WITH MINIMAL ENVIRONMENTAL IMPACT

$$CO_2 + 4H_2 = 2H_2O + \text{Methane (CH}_4)$$

❖ Sources of CO_2:
- Exhaust from fossil fuel
- Exhaust from chemical fuel (e.g. Ethanol)
❖ Source of Hydrogen:
  - From water through electrolysis method

Synthetic Methane Reactors

ADVANTAGES

➢ Ultra pure Oxygen produced during electrolysis for industrial & medical applications

➢ Environmental & economic benefits by use of Millions of tons of undesirable CO_2

➢ HEAT produced from reactors as part of chemical reaction used for steam generation and further production of electricity/H2/O2

➢ Methane fuel used as energy for transportation, fertilizer for agriculture & fuel for diesel & gas turbines

➢ Ongoing research for jet engine propulsion

Source: King Power Corporation

Fig. 3.104: Methane Gas

water when the methane molecule is combust in the atmosphere. The ultimate equation is that electrical energy equals hydrocarbon fuel and there is no other net environmental affect in terms of CO_2, hydrogen, water, and oxygen balance. The production process does not remove any carbon dioxide from the environment, but more importantly, it does not add any carbon dioxide to the environment. The oxygen produced out of the electrolysis method is ultra-pure and can be used in medical and industrial applications. Methane fuel can be used immediately as a fuel for internal combustion, diesel, and gas turbine engines.

Development and application of products, equipment and systems through Green technology will surely conserve the natural environment and resources, reducing the negative impact and minimizing the degradation of environment. It is the right time for us to concentrate on the development of newer green technologies for a safe and healthy tomorrow.

3.14 CONCLUSION

The growth of high technology will move up exponentially due to the current status of technology advancement and the progress made to develop the infrastructure. The requisite infrastructure and expertise will provide the much needed cradle for high technology growth in the country. Previously, it was not so easy to support high technology growth in the absence of requisite infrastructure and expertise in the country. Now the scenario has drastically changed over the last few decades. Presently, India is geared up to achieve excellence in many high technology areas for self-reliance. Similarly, it is time now to revisit many materials and their processing technologies to tap their optimum potential. This demands a joint effort among academia, R&D organizations and industries all over the world. There should be no geographical or political boundaries among nations to lift the quality of human life with the advent of new materials and technologies. This is possible with collaborative effort and joint ventures among various countries to share their technological strengths. These collaborative efforts have an enormous business potential and economic growth. Such efforts will lead to prosperity and a peaceful world which will be a happy global home for all.

REFERENCES

1. Resizing the Future: Military Miniaturisation by JR Wilson, International Defense Review, Jun 97.

2. Duerig T.W, Melton K.N, Stoeckel D., Wayman C.M., Engineering Aspects of Shape Memory Alloys, Butterworth Heinemann Ltd: London, 1990.

3. Mauro Dolce, D. Cardone and R. Marnetto, Implementation and Testing of Passive control Devices based on Shape Memory Alloys, Earthquake Engg. and Structural Dynamics, 2000, Vol. 29, pp. 945-96.

4. J. Holnicki-Szulc and J. Rodellar (eds), Smart Structures., 3.High Technology, Vol. 65.

5. N. Krstulovic-Opara and A.E. Naaman, ACI Structural Journal, March-April 2000, pp. 335-344.

6. Hannant, D.J and Keer, J.G., Autogeneous Healing of Ti Based Sheets, Cement and Concrete Research, Vol. 13, 1983.

7. Sun, G. and Sun, C.T., Bending of Shape Memory Alloy Reinforced Composite Beam, Journal of Materials Science, Vol. 30, No.13, pp. 5750-5754.

8. Cebrowski, Vadm A, Garstka, J. Network-centric Warfare: Its Origin and Future, Proceedings of the Naval Institute, 1998.

9. Mo Jamshidi, Systems of Systems Engineering – Principles and Applications, CRC Press, 2009.

10. Rechtin, E. Systems Architecting: Creating and Building Complex Systems, Prentice Hall, 1997.

11. Yeoh, L.W., Syn, H.B., Lam, C.V. An Enterprise Framework for Developing Command and Control Systems, 17th Annual International Symposium of the International Symposium of the International Council on Systems Engineering, 2007.

12. Yeoh, L.W., Teo, T.L., Lim, H.L., Continuous Systems Development of command, Control and Intelligence Systems, 18th Annual International Symposium of the International Council in Systems Engineering, 2008.

13. Bioinformatics: The foundation of present and future biotechnology, K. K. Tripathi, CURRENT SCIENCE, VOL. 79, NO. 5, 10 SEPTEMBER 2000.

14. Stem Cell Research: Opportunities and Challenges, A Sivathanu Pillai, First International Stem Cell Summit India – 2008, Indian Institute of Technology Chennai, 14 November 2008.

15. Caltech Researchers Create the First Artificial Neural Network Out of DNA, California Institute of Technology, 20 July 2011, http://media.caltech.edu/press_releases/13434

16. The Human Brain Project, Henry Markram, Scientific American June 2012.

17. Nano-sized vaccines, Massachusetts Institute of Technology, Anne Trafton, 22 Feb 2011, http://web.mit.edu/newsoffice/2011/nano-sized-vaccines-0222.html

18. Compendium on Indian Capability on Nano Science and Technology, Macmillan, Editors in Chief, V Rajendran, W Selvamurthy.

19. Introduction to Nanoscience and Nanotechnology, Chris Binns, John Wiley & Sons, 2010

20. Nanoscience and Nanotechnology in Engineering, Vijay K Varadan, A Sivathanu Pillai, Debashish Mukherji, Mayank Dwivedi, Linfeng Chen, World Scientific Publishing, 2010.

21. Nanofuture What''s Next for Nanotechnology, J. Storrs Hall, Manas Publications, 2006.

22. Handbook of NANOSCIENCE, ENGINEERING, and TECHNOLOGY, William A. Goddrad, III, Donald W. Brenner, Sergey E. Lyshevski, Gerald J. lafrate, CRC Press, Taylor and Francis Group.

23. Unbounding the Future: the Nanotechnology Revolution, Erix Drexler, Chris Peterson, with Gayle Pergamil, William Morrow and Company, Inc, 1991.

33. Some challenges in biotechnology, Address at the inauguration of National Conference on 'Biotechnology and National Development: Achievements and Challenges', Central University of Jharkhand, Brambe Campus, Dr. APJ Abdul Kalam.

34. ALGAL BIODIESEL: the next generation biofuel for India, A. K. Bajhaiya, S. K Mandotra, M.R. Suseela, Kiran Toppo, S. Ranade. ASIAN J. EXP. BIOL. SCI. VOL 1(4) 2010: pp 728-739.

35. Microbial Bioremediation of Fuel Oil Hydrocarbons in Marine Environment, Sapna Pavitran et. al., Defence Science Journal, Vol. 56, No. 2, April 2006, pp. 209-224.

36. Introduction to Sensors, National Academic Press, http://www.nap.edu/openbook.php?record_id=4782&page=9

37. Technology Focus, Vol. 11, No.5, October 2003, ISSN: 0971-4413.

38. Technology Guide Principles – Applications – Trends, Hans-Jorg Bullinger (Editor), Springer, 2009.

39. Technologies for Future Armoured System and Strategy for Realisation, A Sivathanu Pillai, International Technology Seminar on Future Main Battle Tank and Future Infantry Combat Vehicle, 22 July 2008.

40. Emerging Technologies in Ocean Warfare, A. Sivathanu Pillai, International Conference on Marine Hydrodynamics, 07 January 2006.

41. Integrated Sensing & Processing for Defence Applications, A. Sivathanu Pillai, Keynote address at International Conference on Intelligent Sensing and Information Processing, 15 December 2005.

42. Convergence of Bio-Info-Nano Technologies, A. Sivathanu Pillai, International Conference on Biomaterials Implant Devices and Tissue Engineering, 06 January 2011.

43. CDCA 2009 C5ISR Collaboration Persistent ISR Panel, Presentation by Terry Simpson, Principal Deputy PEO for Intelligence 2 December 2009.

44. Stem cell technologies Current state Future promise, Presentation by Reeve-Irvine Research Center, Unversity of California, Irvine.

45. Brain tumour stem cells, Angelo L. Vescovi, Rossella Galli and Brent A. Reynolds,

46. Nature Reviews Cancer 6, 425-436 (June 2006).

47. Carbon nanotube filters, A. Srivastava, O. N. Srivastava, S. Talapatra, R. Vajtai and P. M. Ajayan, 1 August 2004; doi:10.1038/nmat1192.

48. Advances in Computing, Communication and Intelligence in Defence, A. Sivathanu Pillai, IEEE International Conference on Computing, Electronics and Electricals, 06 January 2011.

49. Advances in Information Processing, A Sivathanu Pillai, International Conference on Information Science & Applications, 06 February 2010.

50. Dimensions of Robotics & Communication Technologies, A Sivathanu Pillai, International Conference on Emerging Trends in Robotics & Communication Technologies, 03 December 2010.

51. Art, Mind and Brain: A Cognitive Approach to Creativity, Howard Gardner, Basic Books,

52. Expanding Horizons of the Mind Science(s), Editors P.N. Tandon, R.C. Tripathi, N. Srinivasan, Nova Science, New York, 2012.

53. Managing Thought, Mary J Lore, Tata McGraw-Hill Edition, 2010.

54. INS/GPS Technology Trends, George T Schmidt, RTO-EN-SET-064, R&T Organization.

55. Glonass Status and Development Plans, Prof. Dr. Grigory Stupak, Presentation in 5th

Meeting of the International Committee on GNSS, Italy 2010.

56. Indian Satellite Navigation Programme, PK Jain, Presentation in 45th Session of S&T Subcommittee of UN-COPUOS, Vienna, Feb 2008.

57. A Sivathanu Pillai, Advanced Materials & Processing Technologies for Defence, International Symposium on Processing and Fabrication of Advanced Materials, IIT Delhi.

58. Measurement of the neutrino velocity with the OPERA detector in the CNGS beam, T. Adam, N. Agafonova et.al., static.arxiv.org/pdf/1109.4897.pdf.

59. Nuclear Power in India, World Nuclear Association, http://www.world-nuclear.org/info/inf53.html

60. Not Nuclear Fusion But Thorium nuclear power, http://2011nuclearfusion.alternate-healing-science-christian.ca/nuclear_fusion_msr_lftr.html

61. Thorium Reactors, Pure Energy Systems Wiki, http://peswiki.com/index.php/PowerPedia:Thorium_Reactors

62. Nuclear Electric power: Technology, Society, Safety and challenges, Dr. APJ Abdul Kalam, Address during 21st SMIRT Conference, New Delhi, 7 November 2011.

63. Nuclear power is our gateway to a prosperous future, A.P.J Abdul Kalam, Srijan Pal Singh, 06 November 2011, http://www.thehindu.com/opinion/op-ed/article2601471.ece

64. Computers & Internet, http://www.futuretimeline.net/subject/computers-internet.htm#ref6

65. Introduction to Fusion, Culham Centre for Fusion Energy, www.ccfe.ac.uk

66. Materials and Manufacturing Technologies requirement for Cryostat and Vacuum Vessel In-wall shield system of ITER, Bharat Doshi, Institute for Plasma Research, 23 July 2008.

67. ITER Project Outline, F.R. Casci, Wroclaw (PL), 09th June 2009.

68. The ITER Project – the road to fusion power, Jennifer Hay, Bill Spears, ITER Organization, www.iter-nl.nl/files/images/WEC_ITER.pdf

69. Fusion's Missing Pieces, Geoff Brumfiel, Scientific American, June 2012.

70. Launch Vehicle Technology: A Perspective, Prof. APJ Abdul Kalam, Technical Note No. ISRO-TN-21-81.

71. Performance & Cost Effectiveness of ISRO Launchers – Dr. APJ Abdul Kalam, Dr. A Sivathanu Pillai, Technical Note No. ILV/S-TN-01-83 .

72. Future ISRO Launchers - Dr. A Sivathanu Pillai, Report No. ILV/S:TN:32:85/S .

73. Concept Definition and Design of a Single Stage to Orbit Launch Vehicle—Hyperplane, R Gopalaswamy, A Sivathanu Pillai, S Gollakota, P Venugopalan, M Nagarathinam.

74. India in Space: Towards Space Industrialisation Growth, Strategies and Plans by Air Cmde (Retd.) R Gopalaswamy.

75. Solar Electric Power Generation from Outer Space and its Transmission to Earth by Air Cmde (Retd.) R Gopalaswamy.

76. Indian Rockets, Space Yuga 07 March 2011, http://spaceyuga.com/indian-rockets/

77. India announces first manned space mission, Habib Beary, BBC News, 27 Jan 2010, http://news.bbc.co.uk/2/hi/8483787.stm

78. Eleventh Five year Plan (2007-12) proposals for Indian space programme.

79. ISRO plans manned mission to moon in 2014, Bibhu Ranjan Mishra, Business Standard, Oct 08, 2008, http://www.business-standard.com /india/storypage.php?autono=336718&chkFlg

80. India's Space Odyssey - by Raj Chengappa - India Today, pp 60 to 66 - February 5, 2007.

81. Plan panel okays ISRO manned space flight, Priyadarshi Siddhanta, 23 Feb 2009,

http://www.indianexpress.com/news/plan-panel-okays-isro-manned-space-flight/426945/

82. Indian human spaceflight programme, Wikipedia,
     http://en.wikipedia.org/wiki/Indian_human_spaceflight _program

83. Envisioning an Empowered Nation, APJ Abdul Kalam, A Sivathanu Pillai, Tata McGraw Hill, 2004.

84. Space Enterprise - Core Competencies & Low Cost Access, A Sivathanu Pillai, Address at World Space-Biz 2010, Bengaluru, 25 Aug 2010.

85. World Space Vision 2050, APJ Abdul Kalam, Address at the International Aerospace Conference Celebrating Fifty years of Space Technology At California Institute of Technology, Pasadena, California, 20 Sep 2007.

86. 21ˢᵗ Century Dimensions of Space Applications, Dr. APJ Abdul Kalam, Address at the National Conference on Space Transportation System: Opportunities and Challenges, VSSC Thiruvananthapuram, 16 Dec 2011.

87. Hypersonic Technology Demonstrator Vehicle, Presentation by Dr. V. Ramanujachary, DRDL.

88. Geospatial technologies for sustainable development, APJ Abdul Kalam, Address at Geospatial World Forum 2012 Conference, Amsterdam, 24 April 2012.

89. Vision for Space Missions, APJ Abdul Kalam, Symposium on Launch Vehicles: Past, Present and Way Ahead, VSSC Thiruvananthapuram, 28 July 2005.

90. Strategy for Global Competitiveness, A Sivathanu Pillai, Dr. Vikram Sarabhai Memorial Lecture, 13 August 2007.

91. India's Space Effort Beyond Chandrayaan: A Perspective, A. Sivathanu Pillai, International Conference on Emerging Scenarios in Space Technology and Applications, 13 November 2008.

92. Low Cost Access to Space, A. Sivathanu Pillai, Space Expo 2011, 25 August 2010

93. Energy Independence and Sustainable Future, APJ Abdul Kalam, IIT Hyderabad, 21 March 2012.

94. Vision, Regional Cooperation and National Missions, APJ Abdul Kalam, Address at the Global Policy Forum, Yaroslavi, Russia, 10 September 2010.

95. Revolution in Leadership, A Sivathanu Pillai, Pentagon Press, 2011.

96. DRDL Dare Devil Days, Prahlada, AV Rangarao, Defence Research & Development Laboratory.

97. IGMDP Integrated Guided Missile Programme, Defence Research & Development Organisation, 2008.

98. India – An Emerging Strategic Power, A. Sivathanu Pillai, International Institute of Strategic Studies, London, 22 May 2007.

99. Formulation, Execution and Implementation of R&D Projects, A. Sivathanu Pillai, International Seminar on Defence Acquisition, 12 July 2011.

100. Cutting Edge Technologies for Nation's Defence, A Sivathanu Pillai; VK Krishna Menon Memorial Lecture on Science and Technology, 26 December 2009.

101. Indian Ballistic Missile Defence Programme, http://en.wikipedia.org

102. Challenges in Implementation of Defence Projects - Some Experiences, A Sivathanu Pillai, Presentation in 18ᵗʰ Global Symposium 2010.

103. Indian Defence Technology – Aeronautical Technologies, Defence R&D Organisation, Ministry of Defence.

104. Indian Defence Technology – Missile Technologies, Defence R&D Organisation, Ministry

of Defence.

105. India's Armament Programmes, Geopolitics, Vol II, Issue XI, April 2012.

106. C4I2SR - Battle Preparedness for Tomorrow, A Sivathanu Pillai, C4I Asia Conference, Singapore.

107. Hypersonics and Hypersonic Airbreathing Technologies, Marcus Lobbia, Marcus Shaw, The Aerospace Corporation, January 2006.

108. Hypersonic Technology Demonstrator Vehicle, Presentation by Dr. V. Ramanujachary, DRDL.

109. Leadership in Hypersonics Programme, APJ Abdul Kalam, Address during the Inauguration of Centre of Excellence in Hypersonics at Robert Bosch Centre, IISc, Bangalore, 08 November 2011.

110. Hypersonic Reusable Systems for Aerospace and Defence, A Sivathanu Pillai, Keynote Address at second National Symposium on Shock Waves, 27 February 2012.

111. Geoengineering, http://en.wikipedia.org/wiki/Geoengineering

112. Carbon Capture and Sequestration: Potential Environmental Impacts, Paul Johnston, David Santillo, IPCC workshop on carbon dioxide capture and storage.

113. Report on 200 LPH pilot plant trials with industrial wastes during TOT phase, Centre for Fire, Explosive And Environment Safety (CFEES), DRDO.

114. Eco2Green Methane Reactor, King Power Corporation, www.kingpowercorp.com

115. 5G Mobile technologies,
     http://ids.nic.in/Tnl_Jces_May%202012/PDF1/pdf/1.5g_tech.pdf

116. CHAMP high-powered microwaves degrade or destroy electronic targets without collateral damage, http://boeing.mediaroom.com/index.php?s=43&item=2454

PART 4

Technology Spin-offs to Society

When India launched SLV-3 successfully and orbited the Rohini satellite in July 1980, India became the member in the Space Club-It gave me happiness;

When Agni the Intermediate Range Ballistic missile was tested in May 1989 to prove India's might-It added to my happiness;

When India conducted five nuclear experiments in May 1998 and became the Nuclear Weapon State-It gave me further happiness;

When Indian Prime Minister declared that India will become a Developed Nation in 2020, my happiness multiplied;

But when I saw the happy tears rolling out from the eyes of the parents of a polio affected child, on seeing him walking after the fitting of light weight caliper developed from missile technology-IT GAVE ME BLISS

- APJ Abdul Kalam

Cosmology has been of great interest to Science after the Big Bang created the Universe. Insights into the explosion are provided by the remnants of Supernova which is one of the most important sources of energy for the interstellar medium. Supernova shines with the brightness of 10 billion suns with an energy output of 10^{44} joules which is as much as the total output of the sun during its lifetime 10 billion year life time. During this evolution of the universe, galaxies, solar system and the earth, life originated from Amoeba, a single cell organism, to the Human Being, endowed, with six sense organs. The Human Being is a special creation of the creator. Therefore, the life of a human being is highly valuable in all the creations. We are grateful to the creator for giving us this precious life to live. Such a life with its defining structure- the body- has to be free from defects if we have to use the full potential of our birth. As the famous lady poet Avvaiyar says:

அரிது அரிது மானிடராய் பிறத்தல் அரிது;   It is rare of rare to take human birth,

அதனினும் அரிது கூன் குருடு செவிடு   It is rarer than that birth, when the
பேடு நீங்கி பிறத்தல்   human body is free from disability,
                                        blindness, deafness and deformation...

If human life is precious, it is essential to find solution to the defects which enter life by birth or later.

What is that which can help to give a defect free body? The answer lies in technology. The reasoning and the hypothesis of all that is happening lead to Science. When Science finds a product for application, it is technology, which when added with value enhances the prosperity of the society and leads to a better life. Nuclear, Space and Defence technologies may be of strategic importance to mission mode programmes but technologies also have applications to solve problems including healthcare at affordable cost. Imported medical equipment, devices and software solutions for diagnosis, analysis and treatment cost heavily and are unaffordable for many needy people. Therefore, attempt has been made by scientific departments to find out spin-off technologies that serve the society. The succeeding paragraphs describe some of the spin-off technologies developed for the societal missions from the Department of Atomic Energy, Space and Defence Research and Deelopment.

4.1 SPIN-OFFS FROM NUCLEAR TECHNOLOGY

Nuclear energy is one of the cleanest sources of energy. Energy generation is the one aspect of nuclear technology. There are many other applications of nuclear technology to the society, particularly in healthcare. It is obvious that nuclear technology is very important from health and energy point of view. Radio isotopes are used for the diagnostics and treatment of diseases like cancer and thyroid. Coboalt-60 Teletherapy Machine has been indigenously

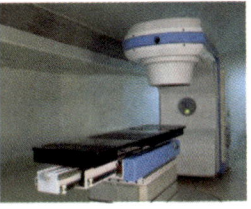

Photo Courtesy:
Technology Innovation Management and
Entrepreneurship Information Services

**Coboalt-60
Teletherapy Machine**

developed by BARC for treatment of cancer. Nuclear technology is being widely used in medical imaging of human body through non-invasive imaging techniques. Diagnostic radiopharmaceuticals used to find the morphology and dynamic functioning of the internal organs and therapeutic radiopharmaceuticals are used for delivering therapeutic doses exactly at the point where the disease exists. The ISOMED facility has been set up by the Department of Atomic Energy to improve the quality of indigenously made healthcare products through radiation sterilization. Robotic assisted surgery helps the patient who needs to be operated by not requiring the Specialist doctors to be physically present. The doctors can perform the operation, sitting in any part of the country/world using a remote. Nuclear Technology has a role to play not only in medicine but also in agriculture. Irradiation of food grains helps in the removal of microorganisms that are the cause of destruction of stored crops and result in spoilage of food. Food irradiation helps in preservation of food for a longer period.

4.2 SPIN-OFFS FROM SPACE TECHNOLOGY

Indian Space technology programmes have led to many societal applications through satellite connectivity in education, medicine, village resource centres, broadcasting, and entertainment for the development of the society. There are also many spin-off products that have emerged out of space 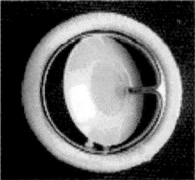 technology, such as Distress Alert Transmitter used for emergency message communication transmission for all type of sea going vessels especially useful for fishermen; Mobile Satellite Service used for communication support when all other means of communication system fails in case of a disaster, Doppler Weather Radar used for precise information about the intensity and radial velocities of cyclones, area rainfall rate and accumulation enabling saving of people much in advance before the calamity happens and so on. Another product is Fire Extinguishing Powder 'OLFEX' for flammable liquid and gas fires and Ternary Eutectic Chloride (TEC) powder for metal fires. In the medical field, the ISRO spin-off product developed from the polyaramid fibres and poly methyl methacrylate is the Artificial Denture Material 'ACRAMID'. Development of 'Chitra heart valve' has become handy to the Thoracic surgeons for replacement of heart valve. This disc contains a tilting disc and a mechanical heart valve prosthesis. Tilting disc is fabricated with ultra high molecular weight polyethylene. The Mechanical heart valve has been customized in four different sizes for aortal and mitral passages in the heart. The valve holds a lot of promise for people suffering from rheumatic heart diseases. TTK Pharma, licensee for the Heart Valve, has launched commercially the 'Chitra Heart Valve', and made it available at an affordable price.

4.3 SPIN-OFFS FROM DEFENCE TECHNOLOGY

The mission of Defence Research and Development Organisation (DRDO) is to design, develop and lead to production of state of the art equipment and weapon systems for the Armed Forces and to achieve self-reliance in critical technologies. Over the last fifty years, DRDO has established itself as a unique organisation with high quality technologists and scientists in different fields ranging from missiles, electronic warfare to underwater systems and materials including advanced composites, processors, supercomputers, robotics, radars and NBC warfare. The technologies developed in realising various equipment and systems have obvious applications to make products which can help the society and industry. Some of the applications have led to affordable quality Healthcare. We took initiative to start spin-off using missile technologies developed by forming Society for Bio Medical Technology (SBMT) integrated with Department of Science and Technology, DRDO and the Ministry of Social Justice and Empowerment. These technology spin-off systems are described below.

Light Weight Calipers (LWC)

LWC was the first product which was designed, developed and successfully produced for the use of poliomyelitis patients to provide the benefits of defence technology spinoffs to society. It is a walking aid for polio patients whose lower limbs are affected. It is made of advanced composites and hence, it is light in weight, durable and cost effective. Conventional caliper weighs three kilograms. The DRDO developed LWC weighing only three hundred grams. Cost of production is exactly half of that of the conventional caliper. It has been fitted in the limbs of more than 40,000 needy children from several parts of India and some of the South East nations.

Orthotically Handicapped – Indian Scenario: Poliomyelitis, or polio, as it is commonly known, is the most common type of physical disability in India. This is a disease that causes paralysis of muscles in childhood if one is not vaccinated with anti-polio vaccine. Essentially, poliomyelitis is a lower limb problem where locomotive functions are affected through the paralysis of quadriceps muscles causing frailty of knee joints which makes a person unstable while walking. In order to stabilize, the person pushes his knee back by the hands, which is referred to as 'hand-on-thigh' gait. Over 60 lakh polio patients in India fall in this category.

Conventional Remedy – Long Leg Caliper: The conventional remedy for paralysed quadriceps is a long leg caliper made of mild steel, leather and wood. This not only weighs excessively but the long pendulum arm rigidises the limb and adds to the difficulties. All the joints are provided in the side bars of the long leg caliper external to the body and are not congruent with the axis of the movement of the natural joints. Apart from technical problems its fabrication is labour intensive and time consuming.

Modern Remedy – Light Weight Calipers (LWC): Dr. PK Sethi of Jaipur

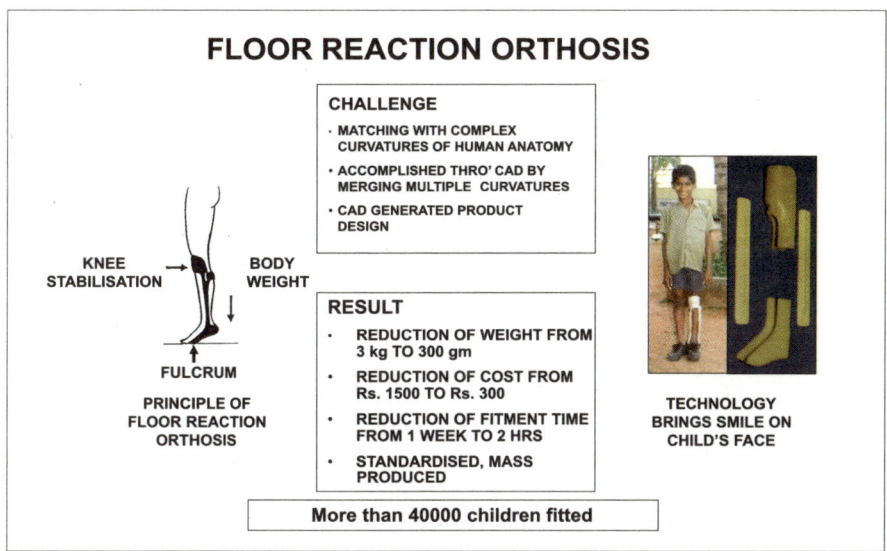

Fig. 4.1: Floor Reaction Orthosis

in collaboration with Prof. SC Lakkad of IIT, Bombay, introduced the tailor-made lower limb calipers, called Floor Reaction Orthosis (FRO), made out of composites/polypropylene in India. The caliper was perfected by them over a period of many years and many poliomyelitis patients were benefitted by these calipers. The success rate and acceptability were quite high among poliomyelitis patients all over the country.

Having mastered the composite material technology and its process, DRDO decided to contribute by providing the required technology support to carry out standardization and production of the LWC. DRDO along with a medical partner, Nizam's Institute of Medical Sciences (NIMS), Hyderabad, led by Dr. BN Prasad and later by Dr. L Narendranath, took up a project for standardization, development and production of LWC. Scientists of DRDO Composites Material Group successfully accomplished this project. E-glass short fibre loading to the extent of 15 to 20 per cent by weight in the polypropylene is found to yield an increase of 20 to 25 per cent in the mechanical properties of the composites and hence is selected for LWC. The process selected is 'injection moulding'- using moulding machine and water-cooled injection moulds for foot plate and knee piece modules. The lateral upright which connects the foot plate to the knee piece was is made out of E-glass reinforced polypropylene sandwich composite by compression moulding process.

Size Standardization of Light Weight Calipers: For standardization, knee and foot zone of a number of polio-affected patients are measured for critical dimensions using an Anthropometric Device (AMD) (Figure 4.2). After due analysis and synthesis of these measurements, using standardisation software packages developed by DRDO, standard sizes for knee and foot pieces are arrived at much like the standard number and sizes we have for footwear. The LWC has mainly three critical parts, they are foot plate, up-rights and knee piece. These three parts form the standard modular FRO. The ready-made modular pieces can be made available off-the-shelf at various rehabilitation/ medical centres. To cover larger population of polio patients, a thigh band is used to connect FRO into long leg caliper. This fitment will only requires thermoplastic welding and riveting operation to assemble the modular pieces, knee joints, side bars and thigh band followed by fine tuning to relieve any pressure points. This fitment and fine tuning will take only a couple of hours drastically cutting down the long waiting time. The fitment will require only one visit of the patient to the rehabilitation/medical centre.

Advantages of the LWC over Conventional Metal and Leather Caliper:

(I)   The cost of LWC is half that of the conventional caliper.
(ii) The weight of LWC (300 gm) is one third of the conventional caliper.

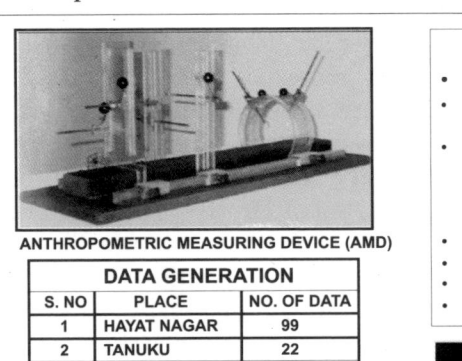

ANTHROPOMETRIC MEASURING DEVICE (AMD)

DATA GENERATION		
S. NO	PLACE	NO. OF DATA
1	HAYAT NAGAR	99
2	TANUKU	22
3	SHAMEERPET	63
4	PUTTAPARTHI	39
5	MH SEC'BAD	16
6	SAROOR NAGAR	102
7	KOLKATA	65
8	BALASORE	15
9	NIMS	200
	TOTAL	621

DESIGN
- DESIGN OF AMD
- DATA GENERATION THROUGH POLIO CAMPS
- FINALISATION OF STANDARD PARAMETERS

'FROSTAN' SOFTWARE
- PARAMETRIC PROGRAM
- ANALYSIS OF DATA
- EVOLUTION OF MODULAR CONCEPT
- STANDARDISATION OF SIZES

MODULAR CONCEPT

FRO

KAFO

Fig. 4.2: FRO Standardisation

(iii) Normal footwear can be used instead of heavy and embarrassing clogs.

(iv) Upon reaching home, one can continue walking or even squatting on the floor with LWC.

(v) The gait resembles that of the normal with LWC, whereas, in conventional caliper, it is very much affected.

Fitment of LWCs: Fitment of LWCs is carried out by technicians under the supervision of an orthopaedic doctor. Polio afflicted children are brought in batches of 20–25 per day for fitment. Local technicians are given training on the fitment of LWCs and thereafter, its maintenance. After the fitment, the children are given gait training for 2-3 days. Medical screening is carried out by local doctors who are also trained for the purpose. The children to be fitted without any corrective surgery are segregated and others are advised for corrective surgeries. The dates of fitment are given with proper medical prescription. The children are reviewed by doctor/orthotist after gait training, and based on satisfactory performance they are relieved/discharged with the fitted LWC's.

Orthotic Knee Joint – A Technology Breakthrough by IIT, Delhi

Light weight long leg calipers have metallic knee joints and side bars. The presence of metal not only makes the LWC heavy but the presence of a metallic joint at the knee also causes a pendulum effect which hinders the normal gait pattern of the polio child fitted with long leg LWC. Also, the cost of a metallic joint is quite high and ranges from Rs.200 to Rs.500 per joint. Therefore, it was felt that there is a need to develop a light weight and durable orthotic knee joint, made out of plastic, for LWCs. A project was given to IIT Delhi for design, development, testing and analysis. Prof. AK Ghosh, Centre for Polymer Science and Engineering and Prof. Naresh Bhatnagar, Dept. of Mechanical Engineering, IIT Delhi undertook this development. The plastic knee joint developed by IIT Delhi (Figure 4.3) and the field trials are under progress at Nizam's Institute of Medical Sciences (NIMS), Hyderabad. So far, the results of the field trials have been satisfactory.

This development is another example where advanced materials' technologies were used for society by making long leg LWC lighter in weight and technically better by avoiding a pendulum effect. This joint was made of a special polymer blend of polycarbonate/acronitrile butadiene styrene/Teflon. It has a

Configuration Design of Universal Joint

• Autonomy of knee joints reveals that torsional, flexural & compressive forces act on it
• Limits of loads (200 Kg with FOS 4) and deflections (0.5 mm at maximum load) finalized
• 06 types of different configuration design prepared
• All the joints having 3 parts with drop lock
• Stress analysis was carried out on all designs
• RPD models were made

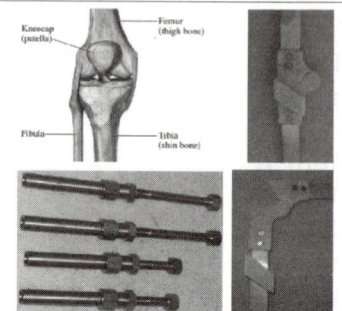

Design of Side Bars

• Limits of loads (200 Kg with FOS 4) and deflections (0.2 mm at maximum load) finalized
• Dimensions were worked for assembly point of view
• Completed configuration design

Design of Pin

• Limits of loads (200 Kg with FOS 4) & deflections (0.05 mm at maximum load) finalized
• Sleeve nut with screw design
• Completed configuration design

Fig. 4.3: Orthotic Knee Joint – Design and Analysis

universal design which combines the features of droplook joint and offset joint in one, whereas, in metallic joints, there are two separate designs for droplook and offset joints. The single universal design of a plastic knee joint has reduced inventory, cost of mould, processing, etc., thereby, reducing the total cost. The design of the plastic joint has been patented. The process for the fabrication of these joints is injection moulding, therefore, a high production rate is achievable. The plastic joint thus developed, weighs only 56 gm vis-à-vis 250 gm weight of a metallic joint. The cost is also likely to be one-fourth of the cost of the metallic joint.

Artificial Limbs

The state-of-the-art, light-weight and cost-effective modular lower limb prosthesis has been designed and developed using carbon-carbon composites. The prototypes developed have been evaluated by Artificial Limb Centre, Pune and Mobility India, Bangalore. The modular design facilitates repair and replacement of individual components. By using various combinations of modular design, it is possible to fit an artificial limb in all amputees above and below knees.

Coronary Stent

The fusion of missile and medical technologies has resulted in the indigenous development of Coronary stent of international standard. The stent is used for maintaining coronary blood flow after angioplasty. Stent is developed from delta-ferrite free matrix of austenitic stainless  steel, largely used in missile components. The stent is fabricated by Andhra Biomedical Components Pvt. Ltd., Hyderabad, for Andhra Cardiology Associates Pvt. Ltd. The biological evaluation of the stent was carried out at Cardiovascular Technology Research Institute and clinically tested at CARE Hospital, Heart Institute, Hyderabad. The indigenous stent is affordable; the cost is nearly one-third of the imported one. Dr. Somaraju used these Kalam-Raju Stents on 7000 patients. Moreover, the availability of indigenous stents in the market has reduced the price of the imported ones. Currently, these stents are going through improvements.

Cardiovascular Catheters

The catheters-based technique offers heart patients the choice of non-surgical treatment of defects within heart and blood vessels. Five types of indigenous catheters developed by DRDO under the aegis of the Society for Bio Medical Technology (SBMT) are used in coronary angiography and for removal of obstruction in blood vessels.

   Biocompatible polymers such as polyethylene (modified), polycarbonate and silicone elastomers were selected for manufacture of catheter components, viz., tubing, hub, sleeve and soft tip. Bio-compatibility of these materials has been established and these catheters have been found to be satisfactory in angiography during clinical trials conducted at LPS Institute of Cardiology (GSVM Medical College), Kanpur. This technology has been transferred to Udhay Kunal Electronics Ltd., for manufacture.

ANAMICA

ANAMICA (ANURAG's Medical Imaging and Characterization Aid) is a DICOM compliant three-dimensional medical visualization software for data obtained from any medical imaging system like MRI, CT and Ultrasound developed by the DRDO Laboratory Advanced

Numerical Research Analysis Group.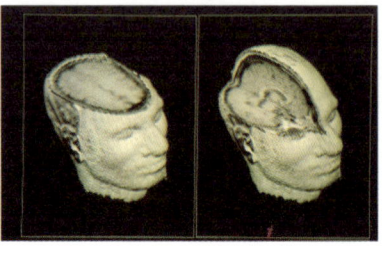
The software has two-dimensional and
three-dimensional visualization
techniques to visualize the images in
various ways. Many three-dimensional
visualization techniques like Iso-Value
surface, Cutout View, Arbitrary Planar
Section etc. are available for the doctor
for visualizing the medical data in three-dimension. The sequence of
images obtained from any imaging system by scanning of a single
patient is packed to form a three-dimensional grid. It is also equipped
with a rich set of image processing and image manipulation
techniqucs. Contrast and Edge Enhancement, Image Algebra,
Pseudo Coloring, Spatial Editing are some of the two dimensional
options provided in the software. The software is also modified for
accepting data from Industrial CT systems. ANAMICA enables the
doctors to simulate the surgery on their computers.

Aspheric Magnifiers

Plastic Aspheric Magnifiers are designed and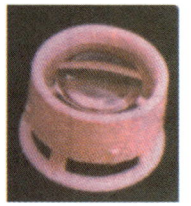
developed by the Instruments Research &
Development Establishment (IRDE) for people
affected with Low Vision. Low vision, at times, cannot
be corrected even by medical or surgical operations or
by using conventional spherical eye glasses. These low
vision aids are lightweight, low cost, impart high field of view and are
available in four powers +16D, +20D, +24D and +28 D. The cost of the
aid is one tenth of the imported cost.

Critical care Ventilator

Fully Functional Critical Care Ventilator, christened INVENTA, has
been designed and developed by PSG College,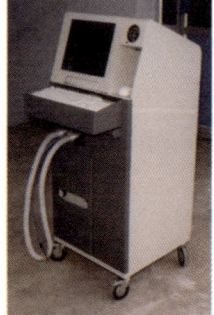
Coimbatore, to meet the needs of Indian
Healthcare after years of extensive research
involving medical professionals, academics and
scientists. The ergonomically designed handle and
castors provide ease of mobility and the unique
tilting provision provided in INVENTA helps in
monitoring the ventilator from many positions in
the ICU. The machine is fitted with unique visual
and audible alarms for identification. The system is

easy to operate with a nine inch wide screen and scroll and select knob. The system takes only a few minutes to start ventilating a patient and all vital parameters are monitored to make effective, fast clinical decisions. The technology has been transferred to M/s. Pricol Limited.

Drishti-1064

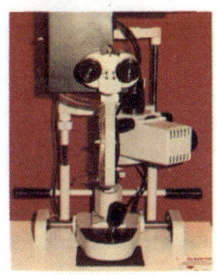

An ophthalmic laser (Nd-Yag) photo disruptor known as Drishti-1064 has been developed under the aegis of SBMT. Drishti is used for capsulotomy and iridotomy. In the former, a cut is made in the membrane formed after cataract surgery to create a central opening thus restoring vision. Iridotomy is a procedure wherein the same laser is used to make a hole in the iris creating an alternate pathway for the outflow of aqueous humor thereby reducing the intra-ocular pressure in cases of glaucoma. The prototype of Drishti-1064 was evaluated by LV Prasad Eye Institute, Hyderabad, by carrying out more than hundred posterior capsulotomy and iridotomy procedures in a span of fifteen months.

The technology has been transferred to Bharat Electronics Ltd., Pune for production. Significant technological expertise in the area of lasers for use in Defence technology has resulted in this high-tech equipment for the common man. The cost of this equipment is reduced to one-third as compared to the imported one.

Titanium Bone Plates, Screws and Dental Implants

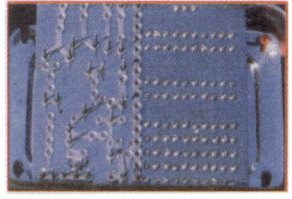

Titanium bone plates and screws have been developed in association with Non-Ferrous Technology Development Centre (NFTDC), Hyderabad, for use in the surgical treatment of facial fractures. Multi-centric clinical trials have been conducted on 130 patients at seven centres belonging to defence and civil sectors. These plates and screws have been found to be suitable for the rehabilitation of fractures of facial skeleton.

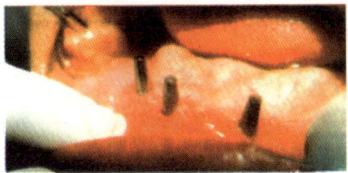

Similarly, titanium mesh of 0.3 mm has been developed for use in cases of jaw fracture and in post oral cancer reconstruction surgery. Yet another product of this association, the Titanium

Dental Implants, has been successfully tested at different centres, viz., Institute of Nuclear Medicine and Allied Sciences, Delhi, King George Medical College, Lucknow and Government Dental College, Ahmedabad.

Orthopaedic Implants and Devices

Total Hip Joint

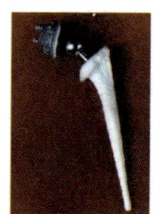

The total hip joint consisting of ball, stem and cup, has been developed in collaboration with the Central Glass and Ceramics Research Institute, Calcutta. The ball is made of alumina; the stem is made of Ti-6AL-4V alloy while the acetabular cap is made of orthopaedic grade ultra high molecular weight polyethylene.

Bone Plates and Bone Screws for Long Bones

The dynamic compression bone plates (6 hole) and screws made of medical grade CP titanium, designed and developed by DRDO and manufactured by Bhavani Engineering Enterprise, Hyderabad, are undergoing multicentric clinical trials at ESI Hospital, Bangalore, Osmania Medical College and Kamineni Hospital, Hyderabad.

Sanjeevani

Sanjeevani is an acoustic detector which can detect any noise under the debris. It is quite useful for the detection of life in the eventuality of natural calamities such as earthquakes, storms, tsunamis, etc.

It consists of a man-portable detection rod fitted with an acoustic transducer which is connected to a monitor where a signal is analyzed. This equipment was extensively used during the Gujarat earthquake and saved lives.

Protection against NBC Radiation

Improvement in radiotherapy of Tumours Using 2-deoxy-D-glucose

The glucose analogue 2-deoxy-D-glucose (2-DG) selectively enhances the radiation damage in tumours by inhibiting cellular repair processes. Phase 1/11 clinical trials have demonstrated the feasibility of administering combined treatment of 2-DG + radiotherapy in patients with brain tumours. Dose optimization clinical trials of Drishti have shown that the combined treatment is well tolerated up to a 2-DG dose of 250 mg/kg body weight without any acute toxicity or significant brain damage.

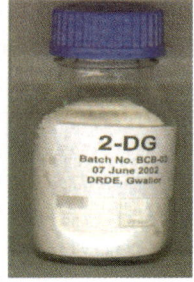

Multicentric trials have been carried out at the Tata Memorial Hospital, Mumbai, Dharamshila Cancer Hospital and Research Centre, New Delhi, and Amala Cancer Hospital, Trissur. The method for the indigenous production of 2-DG has been developed by DRDO. The product has been tested for toxicity and efficacy and found suitable for clinical studies. The transfer of technology has also been effected to Dr. Reddys' Laboratory Ltd., Hyderabad.

Integrated Hood Mask

Integrated Hood Mask (IHM) consists of a mask with a canister and hood. The hood offers additional protection to the wearer against

toxic agents. It is detachable and is used only once. It is made of five panels of three-layered fabric. The outer layer is specially treated and is flame-retardant, water and oil repellent. The front panel is provided with snugly fitting cutouts for anchoring onto the mask. The hood has an adjustable cord with a retainer tab for tightening around the neck, shoulder epaulettes for rank badges and strap at the armpit to prevent the rear portion of the hood from lifting while performing combat related tasks. The light weight and superior quality respiratory mask weighing less than half a kilogram made up of bromo butyle having twin visor and drinking water facility which enables the intake of liquid refreshments. The mask is mounted on a canister and is available in three sizes.

NBC Filter

NBC Filter fat 100m used in the NBC Ventilation systems provides breathable air for the personnel inside the Tank. The filters consist of a particulate and a gas filter in conjunction with filter ventilating units of the vehicles. The particulate filter is cylindrical in shape and assembled over the central core for near removal of all aerosols and toxicant containing harmful particles, bacteria, etc. The gas filter contains impregnated carbon for removal of CW agents like vesicants and nerve agents by physical adsorption. The choking and blood agents are, however, chemically degraded. The other NBC filters fat 200m, 400m and 850m are meant for NBC Protection of the occupants of shelters and ships. All these filters have five years of shelf life in factory packed condition.

Nerve Agent Detector

This equipment works on the electrochemical principle. It consists of an electrochemical sensor connected with a pair of metal wires, a membrane pump and an electrolyte reservoir. The electrolyte in the reservoir is passed through the sensor which contains an organic compound that reacts with the nerve agents to generate cyanide ions. When the wires come into contact with cyanide ions, the potential changes and is taken as a response. The presence of the agents is confirmed by an audio alarm. It is used for the qualitative detection of Nerve Agents like Tabun, Soman, Sarin etc. Salient features of this equipment are instant detection of nerve agents in the field, stability at 0 to 50 deg C and a shelf life of 5 years.

Decontamination Kit and Suit

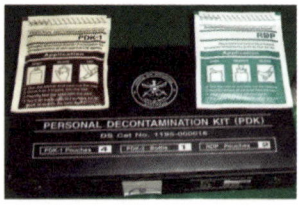

The decontamination suit is an impermeable suit worn over regular garments by an individual engaged in decontamination operations. The suit provides protection against toxic gases, liquid chemical warfare (CW) agents and radioactive dust fallout. The suit can be decontaminated for reuse.

The salient features of this suit are:
*   Complete encapsulation of the wearer.
*   Lightweight nylon fabric coated on both sides with special permeable rubber formulation.
*   Weather resistant as well as flame retardant.
*   Universal size and easy to use.

The decontamination kit has biologically and chemically active substances to deactivate the harmful effect of NBC warfare chemicals/substances on the human body.

Portable Dose Rate Meter (PDRM)

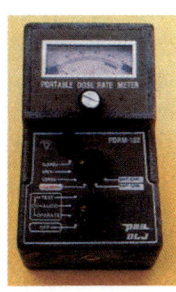

Portable Dose Rate Meter (PDRM) measures X and gamma radiation dose rate in a radioactive contaminated area. This instrument can also be used for emergency operations arising from a nuclear accident or for locating a powerful gamma radiation source. A nylon fabric carrying case with shoulder strap and belt strap for easy handling and storage has also been provided. The PDRM conforms to JSS 55555-L 2 test series specifications.

The salient features of PDRM are:
- Sensor :Energy compensated GM tube
- Range:0 to 1000 R/h in 4 ranges (analog display)
- Accuracy:± 20 per cent
- Power:4 * 1.5 V dry cells (R6/AA size)
- Battery life:> 500 hr for alkaline cells
- Weight:950 g (approx.)
- Size:16 cm * 9.5 cm * 9 cm
- Working temp. :-30°C to + 65°C
- Storage temp. :-40°C to + 75°C
- Relative humidity :Upto 100 per cent
- EMP test: As per MIL-STD

Three-Colour Detector Paper

Three-colour Detector Paper is used for detection of liquid blister and nerve agents. It is a specific dye mixture impregnated paper possessing water repellent property. The paper sticks to any surface and produces three distinct colours when it comes into contact with blister and nerve (G and V) agents. Three-colour detector paper booklet comprises 10 sheets each with easy tearing arrangement. The paper is protected with a plastic cover and can be used under a wide range of temperature and humidity.

The salient features of three-colour detector paper are:

- Size: 135 mm * 70 mm
- Weight: 25 g
- Shelf-life: 2 yr

Auto Injectors

Auto injectors have been specifically
developed for protection against nuclear,
biological and chemical warfare agents. It is a
handy device for immediate self-
administration of atropine and PAM chloride
in the event of exposure to nerve agents.

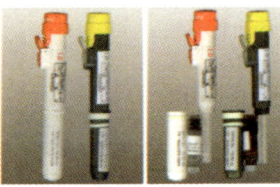

The salient features of auto injectors are:
- Shelf life of Atropine in auto injector is two years at room
  temperature.
- Shelf life of PAM chloride in auto injector is one year at room
  temperature.

Detection Kits

Antibody Detection Kit for Brucellosis

A simple, rapid and inexpensive, dot-ELISA
kit has been developed for detection of
antibodies of Brucella, an organism
primarily afflicting cattle and other domestic
animals. It also affects human beings,
causing pyrexia of unknown origin (PUO)
and low back pain. The animal kit has
undergone field trials by the Department of

Animal Husbandry, Government of Andhra Pradesh, Hyderabad;
Disease Investigation Section, Government of Maharashtra, Pune;
Farah Institute for Research on Goats, Mathura; and Central Military
Veterinary Laboratory, Meerut. The utility of human brucellosis test
kit has been established by trials at Veterinary Biological Research
Institute, Nizam Institute of Medical Sciences and Shree Medical
Centre. The test can be performed in serum, whole blood or milk at
ambient temperature with minor modifications. The cost per test is
only Rs.5.

Typhoid Antigen Detection Kit

This test kit is useful to detect typhoid organisms at an early stage as compared to the Widal test. It is quick, requiring only 8-10 hr in contrast to the conventional blood culture method which requires two to three days to get results. The kit has 95 per cent sensitivity and 100 per cent specificity. The test system is based on rapid growth of the organism in a specially

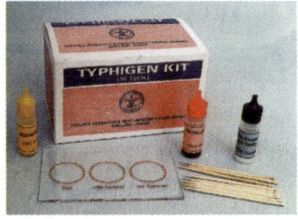

developed medium and only 3 ml of blood is required as compared to 10 ml required for conventional blood culture. The added advantage of this test is that no special equipment or specially trained staff is required. These kits are under production through ToT to an industry.

Leptospirosis Antigen Detection Kit (LEPTODEC)

Leptospirosis Antigen Detection Kit is useful to diagnose leptospirosis disease. Leptospirosis is caused by several serovars of Leptospira. The Dot-ELISA Kit and Sandwich Dot-ELISA kits have been developed for IgM antibody and antigen detection. Dot-ELISA kit provides results equivalent to best available imported kit. It is very simple to use and highly

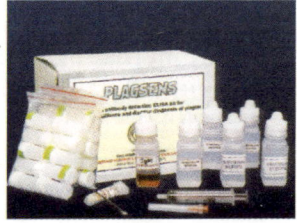

economical. It has been evaluated at National Institute of Communicable Diseases, Delhi; Institute of Microbiology, Madras Medical College, Chennai; and Medical Centre, Cochin. The components of the kit have a storage life of six months at refrigeration temperature. The kit is extremely useful for primary health centres. Sandwich Dot-ELISA kit used for antigen detection from clinical samples of blood and urine. The kit can make confirmatory disease diagnosis on the day of fever itself so as to initiate specific antibiotic therapy. Like the IgM dot-ELISA kit, this is also very simple to use, economical and totally field-based.

Dengue Antibody Detection Kit

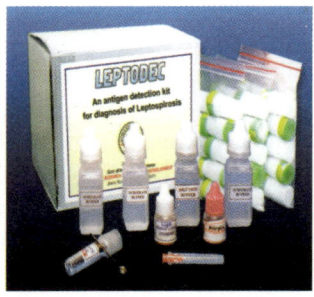

Dengue is the most important mosquito borne viral infection of mankind. Early diagnosis plays an important role in proper control and effective management of patients. A field based Dipstick ELISA kit for the diagnosis of dengue infection enables differentiation of primary and secondary dengue by detecting dengue specific IgM and IgG antibodies in the patient sera. Two identical protocols are employed for detection of IgM and IgG antibodies. However, the specificity of IgM antibody detection was increased by removal of IgG antibodies from patient sera employing Protein A derived from Staph. aureus Cowan I strain. Test can be performed on site and the results are available within three hours. The kit has got a shelf life of more than six months at 4 deg C.

Purified cell culture adapted dengue virus1-4 was used as the antigen in this kit. This kit was evaluated at AIIMS and Hindu Rao Hospital, Delhi, Medical College, Madurai, Medical College, Calicut.

Plague Detection Kit

An Elisa kit for Yersinia Pestis organism identification is required for early detection of plague so that necessary measures are taken immediately to prevent the epidemic. In the kit, rabbit antibodies are coated to the nitrocellulose membrane and the organism. Yersinia Pestis binds to this antibody. The resulting substrate is detected using enzyme substrate reaction. Appearance of brown 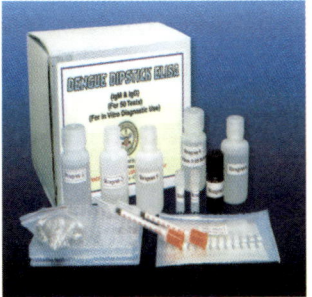 dots indicates positive results. This is suitable for rapid field use. Sample needed is one drop of finger prick drawn blood and the result can be obtained in one hour. The kit has got a shelf life of more than six months at 4 deg C.

Malaria Antigen Detection Kit

Malaria caused by Plasmodium falciparum is a major problem in most of the states of our country. This form of malaria not only causes high morbidity but also leads to high mortality with the manifestation of the dangerous cerebral form of malaria. Plasmodium falciparum synthesizes several proteins containing large amount of amino acid histidine, commonly called as Histidine 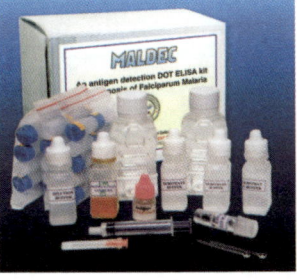 Rich Proteins (HRP). Large amount of HRP II protein is released during the schizont rupture. It is specific to Plasmodium falciparum and has been targeted for detection of acute infection of Plasmodium falciparum.

4.4 CONCLUSION

The benefit of research and development in R&D organisations, academic institutions, and public and private industries should be exploited for societal development, apart from applications in mission mode programmes and projects. Therefore, there is need for collaboration in the area of advanced technologies among the developers, scientific community, doctors, industry and clinical certification agency in order to provide maximum benefit of these technologies to the society. The important factors to fructify such collaborations are: conducive government policies, cost of technology, market potential and competitive pricing. There is also a need to establish an assured market to make production feasible for certain low demand but highly important products. Technology is a tool to relieve pain from society.

REFERENCES

1. Technology Focus, Vol. 10 No. 2 April 2002.
2. Life Sciences Compendium 2008, Defence Research & Development Organisation, Ministry of Defence.
3. Technology Spin-off and Commercialization The case of Dual-use Technologies in India, Manik Kher, Ane Books India, 2007.
4. Spin Offs from Indian Space Program, ISRO, http://www.isro.gov.in/ttg/spinoffs.html.
5. Technology Focus, Vol 18, No. 3 June 2010, ISSN: 0971-4413.

PART 5

Future India

5.1 THREE PHASES OF INDIA'S PROSPERITY

India had gone through a glorious phase two millennium years earlier with remarkable lead in civilisation, culture, knowledge and religious thoughts as well as in education to other nations. That was the *first* phase of India.

Then came the turbulent *second* phase of India, with foreign invasions and dampening of growth path with increased population. Even after the second phase of declining prosperity, India was great in the eyes of the British. Following statement of Lord Mcaulay in the British Parliament on 02 February 1835 proves this point:

"I have travelled across the length and breadth of India and I have not seen one person who is a beggar, who is a thief. Such wealth I have seen in this country, such high moral values, people of such calibre, that I do not think we would ever conquer this country, unless we break the very backbone of this nation, which is her spiritual and cultural heritage, and therefore, I propose that we replace her old and ancient education system, her culture, for if the Indians think that all that is foreign and English is good and greater than their own, they will lose their self esteem, their native culture and they will become what we want them, a truly dominated nation."

The *third* phase started when India became a free nation on 15th August 1947. Now India has large skilled youth, dynamic economy and technological excellence and it is the land of opportunities. The youth have to address issues like poverty, illiteracy, corruption and the temporary economic slowdown. With defined missions in front of us through Technology Vision 2020, why can't India become developed? Can the youths of India pledge to be a part of this transformation? Yes, we are on to our mission of Developed India.

5.2 DEVELOPED INDIA ACTIONS

Distinctive profile of India by 2020 is given in figures that will enable you to take up one or more of the pillars as mission for national development.

To achieve a distinctive profile of India, it is our mission to transform India into a developed nation. As detailed out in Part 3, the five areas identified in the Technology Vision 2020, (i) Agriculture and food processing, (ii) Education and Healthcare, (iii) Information and Communication Technology, (iv) Infrastructure: Reliable and Quality Electric power, Surface transport and Infrastructure for all parts of the country, PURA, and (v) Self reliance in critical technologies are correlated and progress in an integrated way that will lead to overall national progress and development.

India must focus on bringing sustainable development through rural and urban infrastructure, quality education, healthcare, environmental up-gradation, bringing vibrancy in the public institutions for better and enhanced delivery of essential public services on time, reforming the financial system for better global integration and a proactive regulatory system. It is critical to the success of India becoming a Global player. sixty four years of democratic vibrancy to the nation gives confidence to manage the socio-economic turbulences and provide leadership to the one billion plus people in a democratic, multi-cultural, multi-linguistic and multi-religious environment.

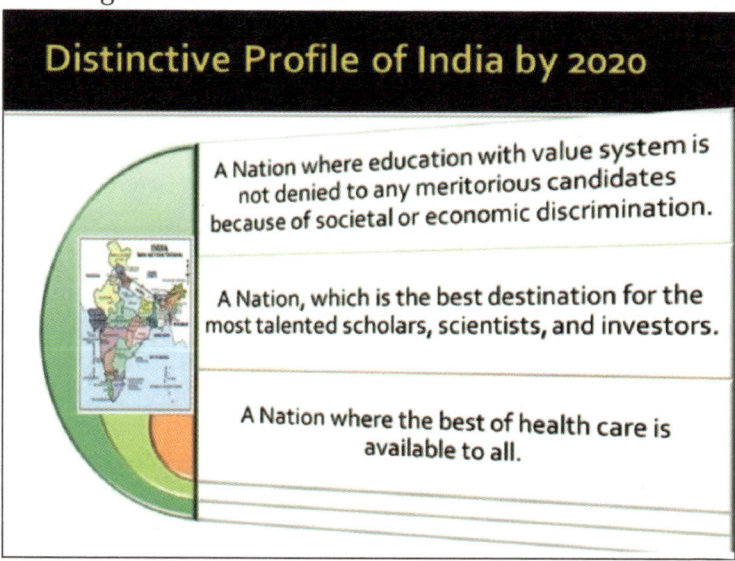

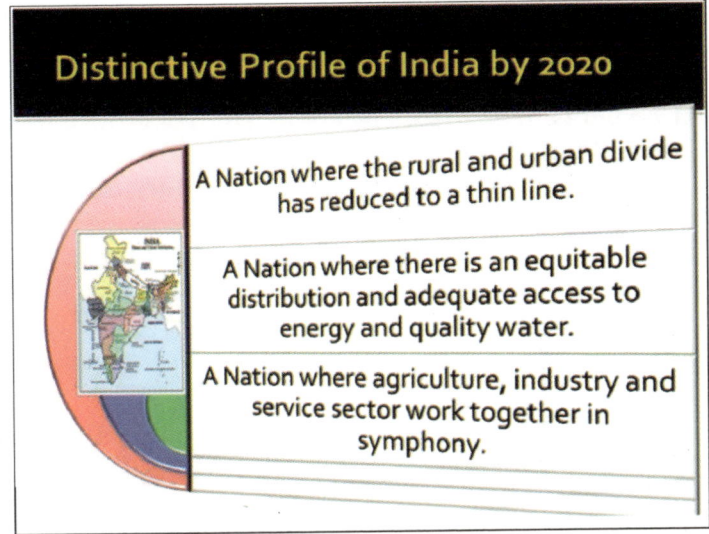

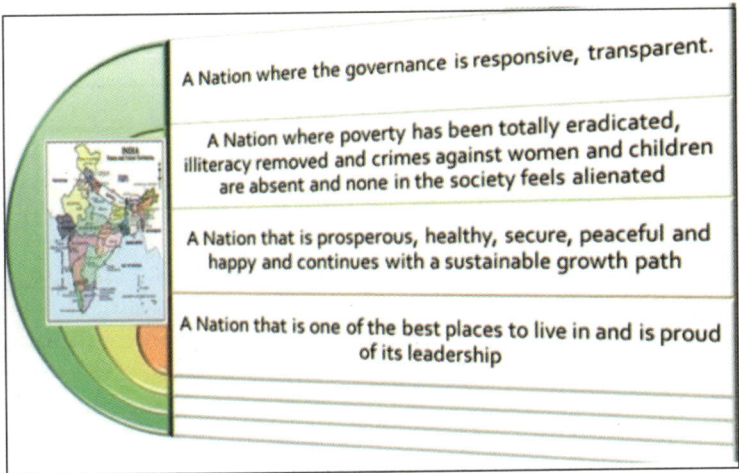

Fig. 5.1: Distinctive Profile of India by 2020

5.3 CURRENT ECONOMIC AMBIENCE OF INDIA

According to World Bank report of March 2012, global economic conditions are fragile. There remain a lot of uncertainties as to how markets will evolve over the medium turn. In the high income countries, the growth is expected at 1.4 per cent in 2012 (-0.3 per cent for Euro countries, 2.1 per cent for the rest). The average growth of developing countries has also got reduced to 5.4 per cent reflecting the growth slowdown. India is presently in the negative direction at around 6 per cent and is expected to go lower in the coming years. But there is great expectation of economy soon bouncing back to a growth of 7-8 per cent by 2015. The comparison of growth pattern for India, China, Brazil, US and Eurozone for 2010 and 2015 and the economy slow down with respect to Agriculture, manufacturing and services are presented in Figure 5.2.

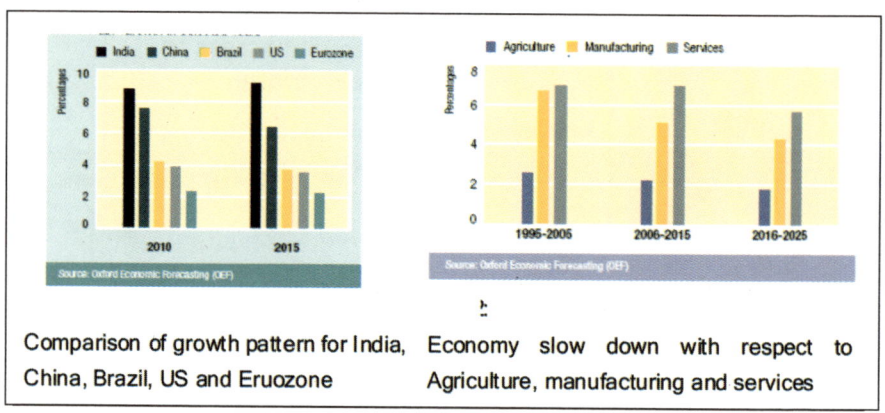

Comparison of growth pattern for India, China, Brazil, US and Eruozone

Economy slow down with respect to Agriculture, manufacturing and services

Fig. 5.2 Comparison of Growth Pattern

The economy in developed countries is disappointing at present and is likely to continue due to the high unemployment situation. In addition, to the above, the turndown in US home prices has further worsened the situation. High income European countries are experiencing negative growth and are in deeper crisis. This is partly because of the intense restructuring that some of these economies are already undergoing. Developing countries also face the economic slowdown but they are not expected to crash to the negative level of growth, as it is seen in the developed countries.

The present top ten largest economies in the world in 2010 in terms of total GDP measured at purchasing power parity (PPP) are the USA, China, Japan, India, Germany, Russia, the United Kingdom (UK), France, Brazil and Italy. India stands at the fourth position. Of the ten, six countries are developed countries. USA accounts for 20.2 per cent and China, the second largest economy accounts for 13.3 per cent of the world total in PPP terms. India, Russia and Brazil are among the emerging economies. These countries have done fairly better than the developed countries during the global economic recession. With the present situation, in another decade there is going to be a major shift in the world economic scenario where the emerging economies will be playing a major role. The rise in importance of emerging economies will have implications for global consumption, investment and environment. Large consumer markets in emerging economies will present enormous opportunities for business. A study by Euromonitor predicts that India, the fourth largest economy in 2010, will overtake Japan to become the world's third largest economy, with GDP accounting for 5.8 percent of the world total in PPP terms if India can overcome the present problems of inflations facing the nation. India could grow even faster due to its younger and faster growing population, provided there is effective management of human resources.

Indian economy was growing at an average of 9 per cent per annum till 2008. In 2009-10 year, Indian economy got affected due to global economic turbulence, but nevertheless it grew at about 7 per cent in 2010-11. Even with the two economic zones USA and Europe looking bleak, India still grew at about 6.9% in the last quarter of 2011. Many financial experts of India felt that the Indian economy is less affected by the world financial crisis.

The reasons are:

(i) The liberalization process in India has its checks and balances consistent with the unique social requirements of the country.

(ii) The Indian banking system has always been conservative which has prevented the crisis.

(iii) The Indian psyche is generally savings oriented and living within means is part of the mindset.

(iv) The 400 million strong middle class structure, with its purchasing power, is indeed providing the stability in the economic structure of the nation. It is a unique economic environment of India and has given inherent strength.

This has reduced the effect of global turbulence on the Indian economy. We will be overcoming the difficulties in the years to come. Sectors like automobile, cement, pharma, ICT and financial services are all posting significant gains. To many Government and business leaders, India seems finally ready to fulfil its destiny as a great nation. To many analysts and economists, India seems set to impact the world as greatly as China has done in the last twenty years.

This is the time innovation has to be encouraged in our thinking to rejuvenate the critical agricultural sector through value addition and to promote the small and medium scale industries for making higher levels of contribution to the GDP through integrated rural development and imaginative products. One can foresee the possibilities of creating new markets through rural potential and employment, giving rise to Public-Private-Citizen partnerships and even International partnership. Government must enunciate policy framework and good tax and duty structure to enable these industries to succeed by capturing higher market share for homemade products as compared to the Chinese products in the market.

5.4 OUT OF BOX IDEAS NEEDED

Out-of-box ideas and innovations are required for achieving the distinctive profile of the nation. We must also remember that challenges faced by the nation are diverse, and yet they are common across different nations and would require cooperation across borders. Technological complexities and cost factors in solving issues like water, energy independence, environment protection, health will require intense cooperation in research, development and operationalization. We need to work together to understand and protect against the fury of nature like earthquakes, cyclones, flood and famine beyond borders. The recent global economic turbulence has taught us the need for a resilient economy. The youth of the nation are ready to contribute to development, and we need to enable them with global cadre skill sets and higher education skills.

5.4.1 Education and Research

Higher Learning in the Past

India was having a famous ancient centre of higher learning from AD 427 to 1197 in Nālandā which is located presently in Bihar State. It was known as "one of the first great universities in recorded history." According to historians, Nalanda flourished in the reign of the Gupta king Śakrāditya (also known as Kumāragupta, AD 415-455) and 1197 CE, supported by patronage from Buddhist emperors like Harsha Vardhan as well as later emperors from the Pala Empire. Nalanda was one of the world's first residential universities, i.e., it had dormitories for students. In its heyday, it accommodated over 10,000 students and 2,000 teachers. The university was considered an architectural masterpiece. Nalanda had eight separate compounds and ten temples, alongwith many other meditation halls and classrooms. There were also lakes and parks. The library was located in a nine-storied building where meticulous copies of texts were produced. The subjects taught at Nalanda covered every field of learning, and it attracted pupils and scholars from Korea, Japan, China, Tibet, Indonesia, Persia and Turkey. But today only the ruins of Nalanda exist. This clearly shows that Indians excelled in knowledge and Nalanda is an apt example. Now, Indian academia is flourishing and we can see a number of talented students passing out every year. The concept of Vasudheva Kutumbakam —"The earth is a family" has now emerged. Utilising this with the advancement of IT, many collaborations could be initiated with foreign universities. Foreign universities may be invited to India to share their expertise and train students here. This will lead to further enhancement of the knowledge base of India.

Changing Education System-Value based Education for All

There is a need to develop human resource with value system and entrepreneurial focus leading to the generation of enlightened citizens. This will need value-based education for all and arresting the tendency of school dropouts. M.R. Raju, Azim Premji, Shiv Nadar and others have developed a model for value education to reduce school dropouts. These models suggest that there is a need for preparing the child for education from three years of age and also providing accelerated learning in school through the use of creative technology. It will help them to develop interest towards learning and make them lifelong learners, not giving up education on trivial grounds. The cumulative effect of all these actions will substantially bring down school dropouts from the existing 35 per cent. With improvement in tele-education technology it is possible to enable quality teaching to reach remote villages through a tele-education delivery system. Also, there is a need to promote education through virtual universities and to create village Panchayat knowledge centres. With globalisation, everyone must concentrate on enhancing their skills so that they become part of a skilled workforce in any part of the world.

Inculcating Creativity

In a competitive world, nations need people with knowledge, skills and resourcefulness. As the society is moving towards knowledge age, the youth needs a first rate value based education system. This must come from the primary school level to develop creative minds. As the nations are becoming more interdependent and opportunities exist for Indian children to move to different countries, the education system must change at the school level to generate interactive creative minds of global cadre.

Cultivating value-based education and creativity will definitely inculcate good R&D work culture among students as they go in for higher studies. R&D is a very important component of any product development. Similarly, multinational companies who are developing high technology products, from different countries, are also looking for a country where they can establish their own R&D house to minimize cost. India is one such location for other countries for establishing their R&D laboratories. Since, India is becoming a knowledge hub, it is an ideal time to utilize the situation to its advantage by generating creativity and R&D work culture among the students to make them suitable for such positions. Indian R&D institutions in biotechnology, nanotechnology, robotics and artificial intelligence, aeronautics, nuclear science, photonics, sensors, smart structures, all will need creative minds for the development of high technology systems and products.

Research–Teaching–Research

University education in India in the present century must be in a competitive environment where research and teaching are integrated. India has a number of IITs, NITs, Universities having arts, commerce, humanities, science, engineering and medical streams. Learning for the students depends upon their aptitude, and creativity of minds to innovate and search for something new. This environment in the educational institutions can be created only by introducing research – teaching – research. Good teaching emanates from research. Teachers love for research and their experience are vital for the growth of institutions and for the students. Any university is judged by the level and extent of the research work it accomplishes. This sets in a regenerative cycle of excellence through quality teaching to the young minds which transform them to researchers. Reform in the education system is therefore essential for high quality human resource.

5.4.2 Co-operative Structure for Agri-production: Processing and Marketing

India's agricultural sector employs about 50 per cent of the workforce, yet accounts for about 17 per cent of total GDP. Last year, the agricultural sector grew at a rate of 2.7 per cent, relative to 11 per cent growth in the services and industry sector. The country has seen considerable successes in Agricultural growth and productivity increase in many parts of the nation. Best insurance to the farmers to increase their earning capacity is possible only by adopting the techniques such as: multi-cropping with regulated drip-irrigation scientific farming, creation of reliable agro and food processing infrastructure, increased availability of ground water and 24x7 power availability. Such a situation is only possible through the creation of responsible agricultural cooperatives which provide collective wisdom as a business house for planning, managing and marketing the agricultural and agro-processed produce for higher level income to the farmer with an assured round-the-year occupation. The essential features of such a cooperative would be:

a) Creation of co-operatives in major agricultural centres involving Government, NGOs, Consultancy Services, R&D Organizations and people as partners.

b) Farmers can be partners in profit by providing farming inputs to cooperative society.

c) R&D Organizations and agricultural universities can provide knowledge inputs for quality agricultural products and technology transfer including training.

d) The co-operative can provide finance, critical inputs provision in time (such as quality seeds, quality fertilizer, and quality pesticides), critical infrastructure, crop insurance and technical guidance such as arrival of monsoon, combating drought etc. to the farmers.

e) The co-operative can be responsible for preservation, processing, diversification, quality assurance, transportation, marketing and profitable operation of the overall business.

f) The Government can provide policy framework for the smooth transparent operation of the co-operative on the lines of the Amul model, Warna model for safeguarding the interest of the farmers and all other stakeholders.

Simultaneously, it will be prudent to create cooperatives for promoting micro enterprises in the non-farm sector for creating profitable non-agricultural business in the rural sector.

5.5 KNOWLEDGE POWER AS CORE COMPETENCY

Knowledge in any form, tacit or explicit, is important. This generates curiosity in young minds and leads to the thirst to know or explore more. Once this thirst is satiated, it leads to further yearning for exploration and this chain reaction leads to knowledge empowerment. Knowledge empowerment ultimately facilitates the attainment of novel goals. The future world would be dominated only by knowledge. The skill set developed through knowledge will redefine the dimension of performance in every sector. In the 21st century, India needs talented youth with higher education for the task of knowledge acquisition, knowledge imparting, knowledge creation and knowledge sharing. Keeping this in mind, our universities and educational systems need to cater to two cadres of youth : (1) a global cadre of skilled youth with specific knowledge for agriculture, manufacturing and services, and (2) another global cadre of youth with higher education for research.

Human resources particularly with a large young population are the unique core strength of the nation. This resource can be transformed into skilled manpower through various educational and training programmes which constitute for creative manpower for wealth generation. Knowledge intensive industries can be generated out of our existing industries by injecting demand for high level software/hardware which would bring tremendous value addition. It is said that "the precious asset of the country is the skill, ingenuity and imagination of its people". With globalization this will become more important because everybody will have access to world class technology, and the key distinguishing feature will be the ability of people to use their imagination to make the best use of technologies."

Nation's Core Competencies

Core competence of India is its large, educated, skilled youth population. This large reserve of human resource is grossly underutilised, in the absence of a nation's development based HR policy. Skilled human resources are being exploited to benefit multinational companies, who provide more opportunities, suitable environment and salary compensation. Hence, it is essential to provide conducive environment and focussed National missions, which will give them the impetus to work for the Nation. This is the only way to boost the indigenous development of high quality, cost effective systems to compete in the global market. Government, academia and R&D organisations need to work together to stimulate and ignite the young minds.

India is popular for its abundant availability of natural resources with significant biodiversity. It is home to 7.6 percent of all mammalian, 12.6 percent of all avian, 6.2 percent of all reptilian and 4.4 percent of all amphibian life including 11.7 percent of all fish, and 6.0 percent of all flowering plant species. India's forest cover ranges from the tropical rainforest of the Andaman Islands, Western Ghats, and North Eastern India to the coniferous forests of the Himalayas; teak-dominated dry deciduous forests of central and southern India; and the babul-dominated thorn forests of the central Deccan and western Gangetic plain. Important Indian trees include the medicinal Neem, widely used in rural Indian herbal remedies. Indian forest cover has many herbs which are of medicinal importance and used in Ayurvedic medicines. Apart from the land, the ocean also has got its own potential. India has a 7,600 km long coastline with nearly two-thirds of the land mass, i.e., 2.02 million sq. km of Exclusive Economic Zone. The Ocean provides many marine species out of which many potential drugs are developed. The Government has initiated a National Project on 'Development of Potential Drugs from the Ocean' by harnessing the potential marine flora and fauna for extraction of drugs for medicinal purposes. The coastal region could be used for water transport and tourism, fishing, water-desalination, deep-sea mining, oil exploration, energy generation through wind and waves, mines and minerals for industrial use and many more. Availability of drinking water and energy through clean sources are the main concern of the day. The Ocean comes to the rescue in addressing these issues. It is the largest and best solar collector on the planet. It can absorb energy equivalent to 250 billion barrels of oil each day. Harnessing 1 per cent of the energy from the Ocean could meet the world's energy needs. A small floating, hexagonal energy Island will harness energy through ocean thermal energy conversion, as well as from wind, sea currents, waves, and the sun. It generates electricity up to 250MW of clean power which is equivalent to one eighth of a large nuclear power plant or one fourth of an average fossil fuel power plant, using the temperature difference between the surface and deep-sea water. Many species, algae and marine sponges are sources for the treatment of many diseases. For example, coral for treatment of HIV & AIDS, marine organisms, shark cartilage for a new cancer drug, chemicals in sponges, snails and algae to treat pain, infections and inflammations, blue green algae for treatment of small-cell lung cancer, marine sponge (Discoderma) used for heart, kidney and liver transplants, horseshoe crab blood used for Intravenous drugs for bacteria, venom of stonefish, box jellyfish and sea snake to produce new and better medicines, etc. Though the effect of the Tsunami was devastating, it brushed aside many mineral materials on the beaches of India. Kerala beach sands have got enormous Titanium raw material and Thorium which could be exploited gainfully.

5.6 WORLD KNOWLEDGE PLATFORM AND NETWORKING OF CORE COMPETENCIES OF THE NATIONS

In the rapidly transforming world of today, there is a definite need for countries to work closely with each other to achieve mutual growth. Any event happening in a remote corner of the world has far reaching impacts on many other nations. No nation can afford to live in isolation and tackle the varying threat perceptions arising every day. It becomes necessary that nations network with each other to exchange resources and inputs to face the world of tomorrow. World Knowledge Platform, networking of the nations will definitely generate employment, free trade, regional common currency and competitive market economy.

5.7 GLOBAL COMPETITIVENESS

Product Excellence and Quality, Cost Effectiveness and Availability in required quantity in time form the three pillars on which competitiveness is built. Once competitiveness is achieved, then the products become world class thus leading the country towards a global leadership. To achieve this unique position what we need is dynamic leadership, creative thinking and innovative approach using technology as the tool.

The building of requisite capability depends on three factors, i.e., design and development, manufacturing and after-manufacturing service support. In the development of critical technologies, the service support becomes an important factor. If this particular area falls in the zone of technology importer, then more efforts are needed. Design and development always falls in the zone of the technology possessor. Therefore, it gives more leverage to have a cost-effective, high quality and superior performance product. Having indigenous R&D capabilities, one will always go for product improvement that is very much required to sustain technology superiority in an environment where technology obsolescence is prevalent. Hence, design and development is the key to a competitive edge (Figure 5.3).

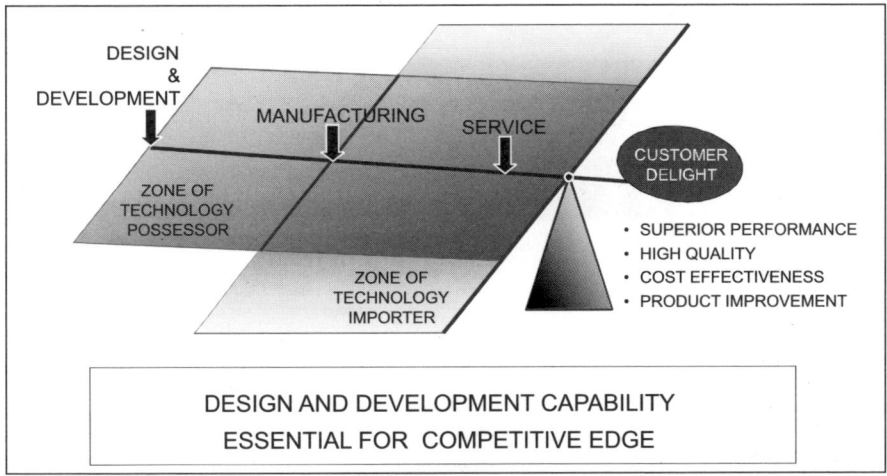

Fig. 5.3: Building Capability

5.7.1 Product Competitiveness by Joint Ventures

Multiple nations can join together along with their core competencies to deliver common development that is beneficial to all. The best example is the unique missile BRAHMOS that was developed jointly by India and Russia. The second example is how networking of African nations (Pan African e-Network) can make progress and contribute to the welfare of the society.

The following paragraphs explain how India used "hard cooperation" with Russia based on its core competence to evolve a world-class product and systems using innovation, creativity, knowledge generation, knowledge sharing, and knowledge dissemination among the scientists of the two countries.

One of the significant technological breakthroughs in India in this decade is the design, development and production of Supersonic Cruise Missile - BRAHMOS by an Indo-Russian joint venture. BRAHMOS is the only supersonic operational cruise missile in the world which can be launched from multiple platforms such as ships, submarines, road mobile and silo, and with modifications from aircraft in multiple maneuvering trajectory profiles and against different types of targets on sea and on land in a network centric warfare. It is a unique weapon system with multi-mission capabilities, unparalleled in the global arena with its speed, precision and devasting power-providing extreme kill energy. This formidable weapon has indeed added tremendous fire power capability to the Indian Armed Forces as distant as 290 km. Most importantly, an

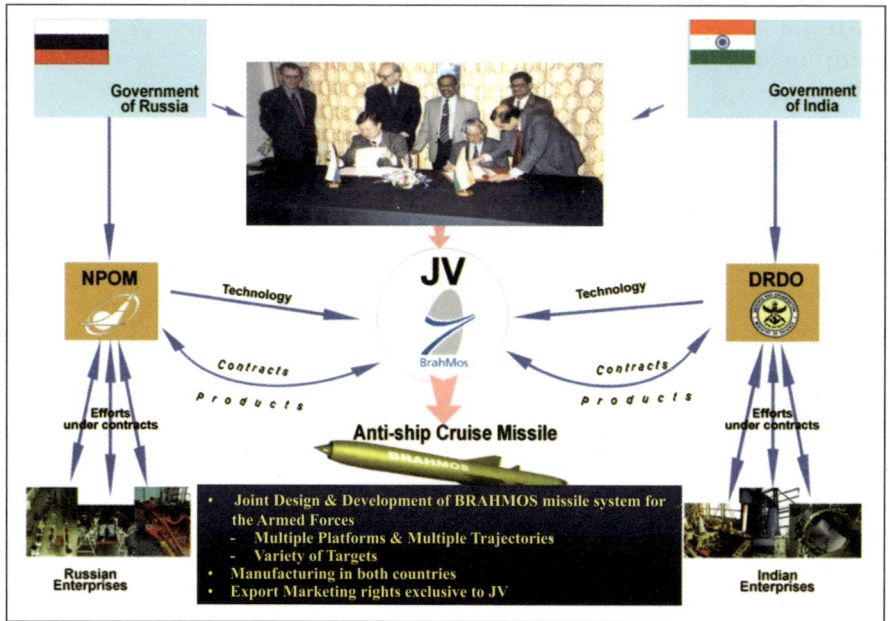

Fig. 5.4: Joint Venture BrahMos

advanced world class system has been realized in the shortest possible
time, leap-frogging other developed countries. This is indeed the
result of the fusion of technologies and scientists from the two
countries with focused mission. In successful design, development,
production and marketing of BRAHMOS missile, an innovative way
of technology co-operation has emerged between India and Russia
generating a business of nearly sven billion dollars with an investment
of 150 million dollars from each partner. BRAHMOS is successfully
inducted in the Armed Forces in multiple platforms, and is also
attracting huge export to friendly countries.

When, in ISRO, we launched SLV3 in 1980, India became the
member of the Space Club and the seventh country to inject a satellite
into orbit. We were proud to achieve this milestone for our country.
When PSLV was launched by ISRO and orbited a satellite around the
moon in 2009, India achieved the special status of the fifth country to
do so. When we were working for strategic missiles in DRDO, the
Intermediate Ballistic Missile Agni was launched in 1989, India
achieved the re-entry capability, we were the sixth nation to do so.
When Agni V was tested in 2012 for a range of 5000 km, India
achieved long range strike capability after the P5 nations. India
became a nuclear weapon state in 1998, and the world called us the

sixth nation to test nuclear weapons. India entered into super computer development two decades ago. At that time, it was told, this will make India the third or fourth nation to have super computer capability. We call this as India going through the fifth country syndrome.

BRAHMOS, the first of its kind in the world, made India the first country to have operational supersonic cruise missile in the Armed Forces, breaking the fifth country syndrome. The Product has established a Brand value and has attracted many countries to demand the system.

BrahMos joint venture is an excellent example of how the core competencies of two countries can be brought together leading to successful enterprises and leap-frog in technology, economic benefits and development of industries. This JV, the first attempt of India with another country to develop and produce a defence product, is a standing role model for others to follow. In this project itself, as a corporate social responsibility, support was given to extend the benefit of the light weight calipers, a spin-off from missile technology. So far, 40,000 polio affected children from India and some ASEAN countries have been benefited.

Ideally, India can design, develop and produce high technology products and systems by joint ventures with developed countries. The pace of technology development is very fast on this route bridging the gap. Gone are the days when we were developing everything in-house. We will be left behind if we follow this route. Thus, technology collaborations can help forward-thinking businesses and organisations accelerate the pace of innovation and bring a competitive advantage in the marketplace. Once the organizations join together, the core competencies of the organisations will synergise and as a result the product becomes competitive in the world market. Thus, innovative technology collaboration will pave the way for societal security.

5.7.2 Pan African e-Network: International Social Responsibility

During the year 2003-04, a proposal was made to create Pan African e-Network for providing seamless and integrated satellite, fibre optics and wireless network connecting 53 African countries. This Indian initiative was accepted by the African Union as the right project to fulfil the needs of African countries.

The Pan-African e-Network project was estimated to cost around US$100 million for India. As part of the project 12 universities (7 from

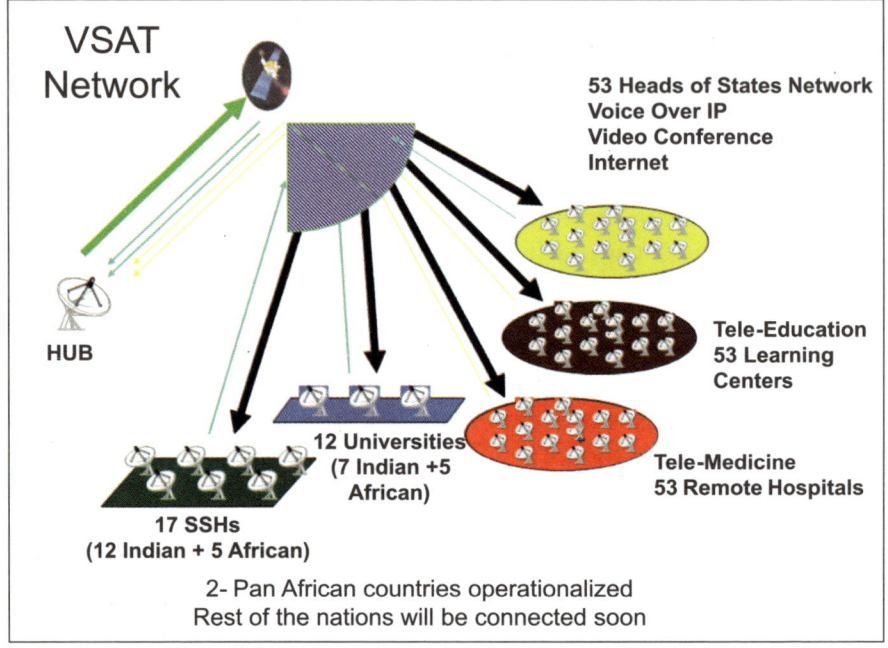

Fig. 5.5: PAN-African e-Network–Overall Architecture

India and 5 from Africa), 17 Super Specialty Hospitals (12 from India and 5 from Africa), 53 tele-medicine centres and 53 tele-education centres in Africa will be connected. The Pan African e-Network primarily provides tele-education, tele-medicine, Internet, videoconferencing and VOIP services. It also supports e-Governance, e-Commerce, infotainment, resource mapping and meteorological services. The pilot project on tele-education and tele-medicine in Ethiopia paved the way for its full fledged deployment. So far 30 countries have been connected and more than 250 sessions of continuing medical education programmes have been taken by 6 Super Speciality Hospitals of India to the doctors of African nations. Two universities have already conducted higher education programmes and completed one MBA course, and continue to provide professional education to the African countries, 6 hospitals are conducting tele-medicine programmes for these countries.

PAN African e-Network is an enabler which has a cascading effect on the socio-economic development of many developing nations and its societies. Enterprises of tomorrow should look at the avenues of bringing about value addition in such enablers which change the environment and rate at which development takes place. Global enterprises may like to facilitate execution of such international social responsibility programs particularly to the needy countries and societies.

5.8 LEADERSHIP TRAITS

To make the country globally competitive with ethics and value system needed for development of the nation which will make and sustain the nation as an economically developed, prosperous, happy and peaceful society, one needs to have creative leadership at all segments of the governance of the nation.

The leadership traits required for a leader are:
i)   Leader must have a vision.
ii)   Leader must have the passion to realize the vision.
iii)   Leader must be able to travel into an unexplored path.
iv)   Leader must know how to manage success and failure.
v)   Leader must have courage to take decisions.
vi)   Leader should have nobility in management.
vii)   Leader should be transparent in every action.
viii) Leader becomes the master of the problem, defeats the problem and succeeds.
ix)   Leader must work with integrity and succeed with integrity.

These essential traits of creative leaders in different areas / fields will definitely make our nation a developed one. What is needed is the spirit among the youth community that "I can do it; we can do it; and the Nation can do it". The educational institutions have got much potential in developing leadership traits among the youth of the nation. The succeeding paragraphs describe the role of leadership in the knowledge society.

5.8.1 Role of Leadership in a Knowledge Society

The world in the 21st century will be a knowledge based society with multiple opportunities naturally in India and we have to become knowledge driven. A book by Denis Waitely, "Empires of the Mind" describes the type of the new world comparing what yesterday was and what today is and specifically emphasizes on what worked yesterday will not work today. Following points give the role of leadership in a knowledge society.

1.   Yesterday - natural resources defined power
     Today - knowledge is power
     Leadership should empower itself with knowledge

2. Yesterday - Hierarchy was the model
Today- synergy is the mandate
Leadership will be the enabler for intersection of multiple faculties towards mission goals

3. Yesterday - leaders commanded and controlled
Today - leaders empower and coach
Leadership will enrich itself through exposure to the needs of sustainable development

4. Yesterday - shareholders came first
Today - customers come first
Leadership should inculcate sensitivity to the needs of all the stakeholders

5. Yesterday - employees took order
Today - teams make decision
Leadership will promote team spirit

6. Yesterday - seniority signified status
Today - creativity drives status
Leadership will be judged by innovation and promote creativity

7. Yesterday - production determined availability
Today - Competitiveness is the key
Leadership will constantly evolve as more competitive with knowledge, management and technology

8. Yesterday - value was extra
Today - value is everything
Leadership will have the priority to inculcate value addition at every level

9. Yesterday - everyone was a competitor
Today - everyone is a customer
Leadership will value feedback and subsequent action based on it.

10. Yesterday - profits were earned through expediency
Today - Work with integrity and succeed with integrity.
Leaders will work with integrity and succeed with integrity and act as promoters of such a culture among their subordinates

5.8.2 Linkage between National Development and Creative Leadership

To keep the aspects of 21st century in mind which will facilitate the Youth of our nation to evolve the learning process for meeting the demands of the 10 components of knowledge society mentioned in section 5.8.1. In order to achieve what is needed is creative leadership. The economic development of any nation is definitely linked with the creative leadership (Figure 5.6).

The linkage between national economic development and creative leadership is as below:

1. A nation's economic development is powered by competitiveness.
2. Competitiveness is powered by knowledge power.
3. Knowledge power is powered by Technology and innovation.
4. Technology and innovation is powered by resource investment.
5. Resource investment is powered by return on Investment.
6. Return on investment is powered by revenue.
7. Revenue is powered by Volume and repeat sales.
8. Volume and repeat sales are powered by customer loyalty.
9. Customer loyalty is powered by Quality and value of products.
10. Quality and value of products are powered by Employee Productivity and innovation.
11. Employee Productivity is powered by Employee Loyalty.
12. Employee Loyalty is powered by employee satisfaction.
13. Employee satisfaction is powered by working environment.
14. Working Environment is powered by management innovation.
15. Management innovation is powered by Creative leadership.

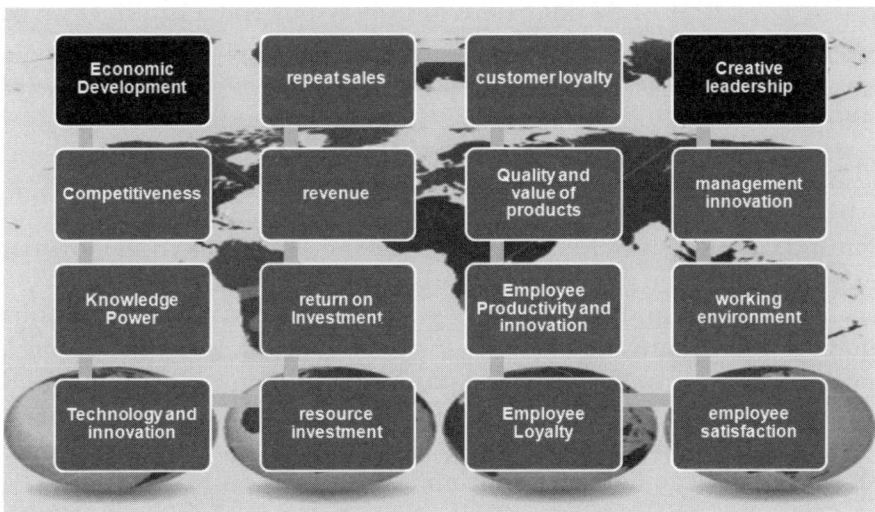

Fig. 5.6: Economic Development and Creative Leadership

For success in all the missions, it is essential to have creative leaders. Creative leadership means exercising the vision to change the traditional role from the commander to the coach, manager to mentor, from director to delegator and from one who demands respect to one who facilitates self-respect. For enhancing enterprise value, we need a large number of creative leaders.

5.9 THOUGHTS FOR CHANGE

Self-reliance is an eternal wealth. The sooner we realize the importance of this wealth it is better for prosperity. Self-reliance provides complete knowledge of a particular matter/system. It unleashes boundless energy and flair to exceed the expectations and goals set by self. It synergizes efforts and entices others to have collaborations and joint ventures. In order to attain excellence, one should strive for self-reliance. Although the broad definition of self-reliance remains the same, its dimensions have changed. In the period of globalization and free market policies, collaboration and joint ventures have become the major strategies of the nations to attain self-reliance in creating technologies and systems. This is the order of the day in the world and likely to prevail in the future too. In the defence sector, India has achieved self-reliance in certain areas and combated Military Technology Control Regimes. The emergence of BRAHMOS has made the entire world look towards India. Export of BRAHMOS will definitely put India in the zone of exporter of defence systems.

A nation's strength predominantly resides in its natural and human resources. In natural resources, India is endowed with a vast coastline and valuable minerals like titanium, uranium, etc. India is also endowed with a rich bio-diversity and human resources, particularly a young population. Knowledge-based value addition to natural resources would mean more earnings for India in the form of exports of finished products instead of merely raw materials. India is a secular country and has got very good diplomatic relations with its neighbours. Be it trade or political or economical, India maintains a positive approach and good relations. The booming economy and strong political relations with other countries have definitely made India a global player. With the heritage of our civilizationand core strengths of large natural and human resources, with value addition and launching of mission projects, the desired goals of food, health and social security, economic prosperity and national security can be

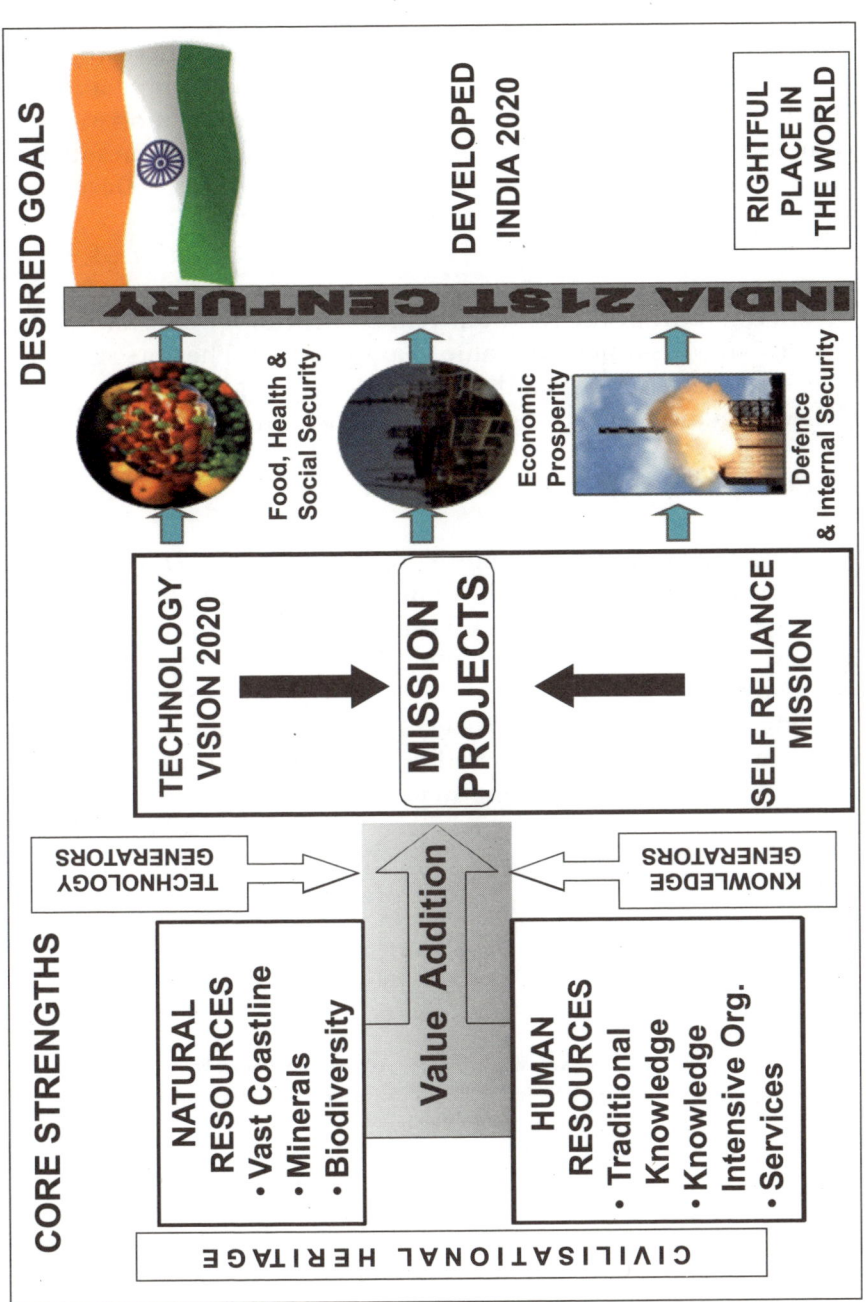

Fig. 5.7: Developed India

achieved leading to a developed India in a decade——An India that will attain its rightful place in the world.

It will regain and retain its rightful place that it has deserved for a long time. It will provide conducive platform to many other nations to collaborate and have cultural, social and business association with it in a completely secure environment in terms of social, economic and national security. This moment will define India as a 'Developed Nation'.

India has to aptly exploit its bountiful resources, knowledge base and nurture the growth of advanced technologies to become a developed nation in the next ten years. Technology and knowledge are the two factors which add value to any product. The core strengths will lead to the desired goal through mission projects, the success of which will make India a strong and self-reliant country.

5.10 CONCLUSION

One must identify one's strengths to exploit them, and must also identify one's weakness to improve upon them for the betterment of life. Keeping this philosophy in mind, many core competencies are aptly being exploited for the benefit of the nation. Integration of the core competencies between the nations will establish world knowledge platform for many technological solutions. Also, new technologies are being researched upon or being obtained through JVs/collaborations to transform them into strengths. The indigenous efforts in many key technologies have made India self-reliant and provided an edge to it. The trend shows that the future India would be a nation which will be the buzzing hub of science and technology.

REFERENCES

1. India 2020 – A Vision for the New Millennium by APJ Abdul Kalam & YS Rajan.
2. Envisioning an Empowered Nation - Technology for Societal Transformation by APJ Abdul Kalam with A. Sivathanu Pillai, Tata McGraw-Hill, 2004.
3. The Scientific Indian, APJ Abdul Kalam, YS Rajan, Viking Penguin, 2010.
4. Report of the Committee on India Vision 2020, Dr. SP Gupta Chairman, Planning Commission December 2002.
5. Profiles of Dynamic Leadership, APJ Abdul Kalam, Valedictory Address Higher Defence Management Course, College of Defence Management, 22 March 2012.
6. Empowering 3 Billion, APJ Abdul Kalam, Address to the students of Indian Institute of Management, Indore, Mar 26 2012,

http://www.abdulkalam.com/kalam/jsp/display_content_front.jsp?menuid=28&menu name=Speeches%20/%20Lectures&linkid=68&linkname=Recent&content=2051&colu mnno=0&starts=0&menu_image=-

7. Economic Development of the nation and evolution of the Policy, APJ Abdul Kalam, Indian Institute of Management, Indore, 25 Mar 2012.

8. Cooperatives for Inclusive Growth, APJ Abdul Kalam, Address at the Inauguration of the International Year of the Cooperatives 2012, NASC Complex, New Delhi 15 May 2012.

9. India Economic Update, March 2012, Economic Policy and Poverty Team, South Asia Region, The World Bank.

10. India and the World:Scenarios to 2025, World Economic Forum.

11. The World in 2050, HSBC, Economics Global, 04 January 2011.

12. The World is Flat: A Brief History of the Globalized World in the 21st Century, Thomas Friedman, Penguin Books, 2005.

Overall Conclusion

Technology Influence

The future of the world is going to be completely dominated by technology. In our daily life, we find ourselves so dependent on technology and this dependence is likely to increase manifold in times to come. Technology is not only the synonym of advancement but also a way to live life, which should be endowed with social security, economic security and health security. The key to these securities lies in technology. In the modern world, the expectancy of life has increased, which was a distant dream in the past. Many new inventions and the apt use of technology saw through many healthcare measures which are cost effective, so that a common man can afford it. Moreover, the quality of healthcare has improved drastically due to technology intervention. The military might of a nation is directly proportional to its technological excellence which ensures national security and economic and social security. But the relationship of various securities and technology is not all that simple. It requires high quality research input. Development of high technology and later its translation into high quality and reliable systems takes years of hard work by a group of experts. Development and change are constant processes and one should keep striving for betterment. The inquisitiveness, developed through research activity, propels one to strive further. Inquisitiveness is the fuel for research. High quality research leads to development of high quality and reliable systems, which make a country developed and advanced.

Science Is Indeed Reciprocating

The growth of science and technology has been phenomenal in the recent past, improving the quality of life of the Human Being. The emergence of new technologies has opened up myriad of applications and now it is left to the ingenuity and imagination of the human mind to explore and exploit them further. With the established base of multiple technologies all over the world, scientific minds have to come together for new innovations and increasing efficiency for energy independence, access to drinking water, curing of dreaded diseases, clean and green environment, safe living, so that there is happiness everywhere and billions of people can smile. Core competencies of the nations can be integrated to form World Knowledge Platforms with technology driven mission mode programmes for achieving the above need.

Taking into account India's specific requirements of value based education and skill development of the large youth power and in order to focus the energy in the direction of development for achieving global leadership for India, ten unique technologies have been discussed. These technologies are multi-disciplinary, amalgamating various disciplines of science and engineering leading to new products and capabilities. This provides an opportunity to the youth to elevate to a higher level of knowledge and to think differently from the yesteryear generation leading to innovation of high performance products. This will result in the competitiveness of the Nation in the globalised world.

There are pessimists in India who ask the question whether India can be a developed nation and a global leader, citing economy slow down, brain drain, poverty and so on. We need to become Captains to solve these problems. Our experiences in developing crucial technologies for missile programme overcoming technology denials from the developed world, in evolving Control Law for Light Combat Aircraft through a National Team in nine months time, inspite of the unilateral termination of contract by an American firm due to May 1998 Nuclear experiments, and in evolving a unique joint venture BrahMos to develop the world's best supersonic cruise missile taking global leadership by harnessing the core competencies of two countries, confirm that India can be on par with the other developed nations. The path is known. Taking technology leadership with innovation is the need of the hour. The youth of India have to rise to the occasion, utilise the opportunities and prove that India can be a global leader.

India is blessed with the largest youth power in the world. Skill and value system will make India a globally demanded human resource. Ignited mind of the youth is the most powerful resource on the earth, above the earth and under the earth. Their knowledge, courage and devotion will certainly make India a great nation, once again!

Courage to think different
Courage to invent
Courage to travel into unexplored path
Courage to discover the impossible
Courage to combat the problems and succeed
Are the unique qualities of the Youth

Abbreviations

AAD	Advanced Air Defence
AHWR	Advanced Heavy Water Reactor
AIDS	Acquired Immuno Deficiency Syndrome
AMD	Anthropo Metric Device
AND	Ammonium Di Nitramide
ASLV	Augmented Satellite Launch Vehicle
AVATAR	Aerobic Vehicle for Hypersonic Aerospace Transportation
BARC	Bhabha Atomic Research Centre
BMD	Ballistic Missile Defence
BPO	Business Process Outsourcing
C4I2SR	Command, Control, Communications, Computer, Information, Intelligence, Surveillance and Reconnaissance
CD	Compact Disc
C-DAC	Centre for Development of Advanced Computing
CFC	Carbon Fibre reinforced composites
CFD	Computational Fluid Dynamics
CFL	Compact Fluorescent Lamp
CNT	Carbon Nano Tube
CO_2	Carbon dioxide
COE	Centre for Excellence
CRO	Clinical Research Organisations
CW	Chemical Warfare
DAE	Department of Atomic Energy
DB	Double Base
DEW	Directed Energy Weapons
DMB	Digital Multimedia Broadcasting

DNA	Deoxyribo-Nucleic Acid
DRDL	Defence Research and Development Laboratory
DRDO	Defence Research and Development Organisation
DVD	Digital Video Disc
ECCM	Electronics Counter Counter Measures
EOCM	Electro Optical Counter Measures
ERA	Explosive Reactive Armour
ESD	Electrostatic discharge
EU	European Union
FADEC	Full Authority Digital Engine Control System
FBR	Fast Breeder Reactor
FBTR	Fast Breeder Test Reactor
FOG	Fibre-Optic gyros
FRO	Floor Reaction Orthosis
FRP	Fibre Reinforced Plastic
FSAPDS	Fin Stabilised Armour Piercing Discarding Sabot
GaAs	Gallium Arsenide
GDP	Gross Domestic Product
GEO	Geosynchronous Equatorial Orbit
GIS	Geographic Information System
GNCST	Global Net Centric Surveillance and Targeting
GP	Gun Propellants
GPC	Gel Permeation Chromatography
GPS	Global Positioning System
GRB	Gamma Ray Burst
GSLV	Geosynchronous Satellite Launch Vehicle
GTO	Geosynchronous Transfer Orbit
HAL	Hindustan Aeronautics Ltd
HEAT	High Energy Anti Tank
HIV	Human immunodeficiency virus
HLLV	Heavy Lift Launch Vehicles
HPL	High Power Lasers
HPM	High Power Microwaves
HRP	Histidine Rich Proteins
HSTDV	Hypersonic Technology Demonstrator Vehicle
HTPB	Hydroxyl-Terminated Poly Butadiene
IaaS	Infrastructure as a Service
ICAL	Iron Calorimeter
ICT	Information and Communication Technology

IED	Improvised Explosive Device
IFF	integrated identification of friend or foe
IGMDP	Integrated Guided Missile development programme
IGS	Inertial Guidance System
IHM	Integrated Hood Mask
IIR	Imaging Infra-Red
IISc	Indian Institute of Science
IIT	Indian Institute of Technology
INCOSPAR	Indian National Committee on Space Research
INO	India-based Neutrino Observatory
INSAT	Indian National Satellite
IRBM	Intermediate Range Ballistic Missile
IRS	Indian Remote Sensing satellite
ISF	International Space Force
ISR	Intelligence, surveillance and reconnaissance
ISRO	Indian Space Research Organisation
IT	Information Technology
ITER	International Thermonuclear Experimental Reactor
ITeS	Information Technology Enabled Services
JV	Joint Venture
KDCMPUL	Kaira District Cooperative Milk Producers Union Limited
KE	Kinetic Energy
LCA	Light Combat Aircraft
LCC	Launch Control Centres
LED	Light Emitting Diodes
LEO	Low Earth Orbit
LEPTODEC	Leptospirosis Antigen Detection Kit
LFTR	Liquid Fluoride Thorium Reactor
LOBL	Lock On Before Launch
LOVA	Low Vulnerability
LRTR	Long Range Tracking Radar
LWC	Light Weight Callipers
MBR	Multi Barrel Rocket
MBT	Main Battle Tank
MCC	Mission Control Centre
MEMS	Micro-Electro-Mechanical Systems
MEO	Medium Earth Orbit

MFCR	Multifunction Fire Control Radar
MIP	Moon Impact Probe
MMW	Milli Metric Wave
MOEMS	Micro-Opto-Electro-Mechanical Systems
MSR	Molten Salt Reactor
MTCR	Missile Technology Control Regime
NAMICA	Nag Missile Carrier
NASA	**National Aeronautics and Space Administration**
NDDB	National Dairy Development Board
NFTDC	Non-Ferrous Technology Development Centre
NGO	Non-governmental organization
NIMS	Nizam's Institute of Medical Sciences
NSG	Nuclear Supplier Group
NSTI	Nano Science and Technology Initiative
OBC	On-Board Computer
OPERA	Oscillation Project with Emulsion-tRacking Apparatus
OV	Orbital Vehicle
PaaS	Platform as a Service
PAD	Prithvi Air Defence
PBPCs	Peripheral Blood Progenitor Cells
PDRM	Portable Dose Rate Meter
PPB	Parts per Billion
PPP	Purchasing Power Parity
PSLV	Polar Satellite Launch Vehicle
PUO	Pyrexia of Unknown Origin
PURA	Providing Urban Amenities in Rural Areas
R&D	Research & Development
RCI	Research Centre Imarat
RLG	Ring-Laser gyros
RLV	Reusable Launch Vehicles
RPC	Resistive Plate Chambers
SaaS	Software as a Service
SBMT	Society for Bio Medical Technology
SDSC	Satish Dhawan Space Centre
SLM	Spatial Light Modulator
SLV	Satellite Launch Vehicle
SMA	Shape Memory Alloys
SPS	Solar Power Satellite
SRE	Space Capsule Recovery Experiment

SSTO	Single Stage to Orbit
TB	Triple Base
TEC	Ternary Eutectic Chloride
TERLS	Thumba Equatorial Rocket Launching Station
TIFAC	Technology Information, Forecasting and Assessment Centre
TSTO	Two Stage to Orbit
UAV	Unmanned Aerial Vehicles
UGV	Unmanned Ground Vehicle
UV	Ultra-Violet
VRC	Village Resource Centres
WWW	World Wide Web

Index